STATISTICAL EXPLORATIONS

WITH MICROSOFT® EXCEL

STATISTICAL

EXPLORATIONS

WITH

MICROSOFT® EXCEL

Millianne Lehmann
Paul Zeitz

University of San Francisco

Duxbury Press

An Imprint of Brooks/Cole Publishing Company

I(T)P® An International Thomson Publishing Company

Pacific Grove • Albany • Belmont • Bonn • Boston • Cincinnati • Detroit • Johannesburg • London
Madrid • Melbourne • Mexico City • New York • Paris • Singapore • Tokyo • Toronto • Washington

Sponsoring Editor: Curt Hinrichs
Project Development Editor: Cynthia Mazow
Marketing Team: Marcy Perman, Michele Mootz
Editorial Assistant: Rita Jaramillo
Production Editor: Timothy Wardell

Manuscript Editor: Caroline Jumper
Permissions Editor: Carline Haga
Cover Design: Kelly Shoemaker
Cover Printing: Malloy Lithographing
Printing and Binding: Malloy Lithographing

For more information, contact Duxbury Press at Brooks/Cole Publishing Company:
511 Forest Lodge Road, Pacific Grove, CA 93950, USA

International Thomson Publishing Europe
Berkshire House 168–173
High Holborn
London WC1V 7AA
England

International Thomson Publishing GmbH
Königswinterer Strasse 418
53227 Bonn
Germany

Thomas Nelson Australia
102 Dodds Street
South Melbourne, 3205
Victoria, Australia

International Thomson Publishing Asia
221 Henderson Road
#05–10 Henderson Building
Singapore 0315

Nelson Canada
1120 Birchmount Road
Scarborough, Ontario
Canada M1K 5G4

International Thomson Publishing Japan
Hirakawacho Kyowa Building, 3F
2-2-1 Hirakawacho
Chiyoda-ku, Tokyo 102
Japan

International Thomson Editores
Seneca 53
Col. Polanco
11560 México, D. F., México

Printed in the United States of America.
10 9 8 7 6 5 4 3 2

Library of Congress Cataloging-in-Publication Data

Lehmann, Millianne.
 Statistical explorations with Excel/Millianne Lehmann and Paul Zeitz.
     p.   cm.
 Includes bibliographical references and index.
 ISBN 0 534-51611-4
 1. Statistics—Data processing. 2. Microsoft Excel (Computer file). I. Zeitz, Paul, 1958-. II. Title.
QA276.4.L455 1998
001.4'22'02855369—dc21   97-18069
                                                                   CIP

Contents

Detailed Contents

Chapter #3 ?
mean, median, range,
variance, standard deviation

#2   Chapter #2.

chapter #6 in textbook.

chapter #7

chapter #4 in text book.

Preface

Statistical Explorations with Microsoft Excel provides Excel-based activities designed to help students to organize and analyze real data sets and to investigate fundamental statistical concepts.

The book is not in itself a statistics text nor is it a complete Excel manual. There are several excellent books on the market which provide a comprehensive guide to Excel's statistical capabilities.[1] This book is no such compendium. It is designed to be used with any introductory statistics text, and it is written to be read at the computer while working through the various Excel "recipes."

The documentation for these activities is detailed and lavishly illustrated so as to minimize (unfortunately, not eliminate) computer frustration. We have found that there is no such thing as too much documentation. For example, you cannot simply write "copy the contents of range so-and-so from worksheet one to worksheet two." Instead, each mouse click and every single press-and-drag must be specifically described in a detailed "recipe"; otherwise, the typical student is soon at sea and terribly frustrated. To the experienced Excel user such instructions may seem annoyingly detailed and repetitive. For the typical computer novice they are definitely not.

The material in *Statistical Explorations with Microsoft Excel* was developed for use in a general education statistics course required of all students at the University of San Francisco (USF). The typical student enrolled in the course is a freshman Liberal Arts major whose academic interests tend to be non-mathematical and whose experience with computers has been limited, by and large, to word processing. Many students in the course would never have elected to take statistics were it not a graduation requirement and will make no professional use of statistics. "For such students statistics is not a technical tool but part of the general intellectual culture that educated people share."[2] Hence the requirement. The book is also used in the honors sections of the general education statistics course where most students are business or science majors whose mathematical aptitude and computing backgound is somewhat more sophisticated and whose professional need for training in statistics is unquestioned.

---

[1]For example, Michael R. Middleton, *Data Analysis Using Microsoft Excel*, Duxbury Press, 1997
[2]David S. Moore, *Statistics, Concepts and Controversies*, W.H. Freeman, 1997, pg. x

An important goal for all of these students is a competence with Excel sufficient for the analysis of moderately large, moderately complex data sets using charts, histograms, boxplots, scatter plots, five-number summaries, means, medians, modes, and standard deviations. A second and equally important application of computing is the investigation of fundamental statistical concepts such as the Central Limit Theorem, the meaning of statistical significance, and the interpretation of confidence intervals through Excel-based activities and explorations. An added benefit for many students is the marketability of the Excel skills gained.

Why Microsoft Excel?

First of all, Excel provides an accessible statistical package which is more than adequate for an introductory treatment of statistical ideas.

Second, spreadsheet skills have a general applicability only slightly less important than that of word processing. Since our general education program at USF takes computer literacy as one of its objectives, Excel's status as the premier spreadsheet package was decisive.

Students who will make professional use of statistical methods will go on to take advanced courses featuring whatever statistical packages are appropriate to their discipline. Sometimes this is Excel, sometimes not. In either case the advanced course builds on the experience with statistical concepts and with computing gained in the introductory general education course.

How to Use this Book

This book has been successfully class-tested as a companion to the following introductory statistical reasoning texts:

> Brase & Brase, *Understandable Statistics*
> David A. Freeman, et al., *Statistics*
> David S. Moore, *Statistics: Concepts and Controversies*
> David S. Moore & George P. McCabe, *Introduction to the Practice of Statistics*
> Jessica Utts, *Seeing Through Statistics*

The book is an appropriate lab manual for use with many other introductory texts including Robert Johnson's, *Introductory Statistics*, 7th. ed., Mason et al.'s *Statistics: An Introduction*, 5th ed., or Larry J. Kitchen's, *Exploring Statistics*, 2nd. ed., to name a few.

The first six chapters of the book provide an introduction to almost all of the Excel features used in the remaining chapters. These six chapters, with the exception of Chapter 4 which can be read at any point after Chapter 3, are written to be completed in sequence, although there are a few optional sections which can be skipped. Chapters 7 through 15, with one small exception, are independent as far as Excel skills are concerned. Section 8.1, which covers the RAND function and Excel's calculation options, is a prerequisite for several of the later chapters. We were careful to write section 8.1 so that it can stand alone independently of the remainder of Chapter 8, and the reader is referred to it when necessary.

Excel skills and features are introduced on a "need to know" basis. For example, we use menu commands almost exclusively even when keyboard equivalents exist. The menu commands are easier to remember and can be searched for if forgotten; the short cuts can be acquired later with experience.

The first one or two exercises in each problem set usually ask the reader to follow the steps outlined in the chapter and print the results. Subsequent exercises apply the techniques to additional data sets and ask for analysis and comment. Many of these later exercises are suitable for class presentation by students working either alone or in groups or by the instructor.

Most chapters have an accompanying Excel workbook which contains the data sets used in the chapter and in the exercises that follow the chapter. These workbooks are found on the diskette entitled **STAT** which accompanies this book.

Students with a background in computing or those more strongly motivated to learn Excel and/or statistics move quickly through the introductory material and complete many of the later chapters. At USF, these students are usually enrolled in the honors sections. The non-honors course, for example, typically does not treat the calculation of p-values or any other aspect of hypothesis testing while the honors sections do.

What Version of Excel?

We assume the reader is working on a Windows machine using either Excel 7 or 97 or on a Macintosh computer using Excel 5. Almost all of the version-specific illustrations are screen dumps taken from Excel 7 or 97. When procedures differ somewhat between these two versions, we have provided two sets of instructions with illustrations in the body of the chapter. The Excel 97 versions of Chart Wizard and Pivot Wizard are substantially different from those of Excel 7 and 5, and so instructions for the Excel 97 versions have been included in appendices.

The dialog boxes in the Macintosh version of Excel differ only superficially, for the most part, from those for the Windows version of Excel 7. Screen dumps unique to the Macintosh have been included only in Chapters 1 and 2 where we show Macintosh Excel toolbars, toolbar buttons, the Excel Window, and the Print dialog box.

STAT: The Computer Diskette

The computer diskette accompanying the book contains the data workbooks and Excel add-ins mentioned earlier. (The diskette is formatted for a Windows (PC) machine. The same diskette formatted for the Macintosh can be obtained from Duxbury Press. See B.4 for instructions.) These workbooks are named after the chapter they accompany, for example, **datach3.xls** is the workbook for Chapter 3. The disk also includes the files for three Excel add-ins which are used in many of the exercises. Two of these add-ins, **Boxplot** and **Smart Histogram**, automate the charting of histograms and boxplots and the third, **SRS**, automates the selection of random samples from a data range. These add-in files must be saved to Excel's Macro Library. Analysis ToolPak, which is an Excel option, is also necessary for many of the chapters and must be added in. Instructions for loading all of these add-ins are given in Appendix B.

Solution Manual

A solution manual is available on diskette from Duxbury Press.

Acknowledgements

We are grateful for the support and assistance of our colleagues, many of whom contributed drafts for several of the chapters and who patiently class-tested this material through several revisions: John Fay, John Kao, Paul Lorton, James Matthews, Tristan Needham, Peter Pacheco, Moe Sadaoui, Jean Simutis, Tom Stillman, and Robert Wolf. We especially thank Professor Needham for his pioneering work on the histogram add-in.

We are indebted to our Dean, Stanley Nel, for his commitment to computer literacy for all students, without which neither this nor many other worthy projects could have gone forward. We thank Ben Baab, Executive Director of Information Technology Services, and his staff for their hours and hours of expert computer support. We are very lucky to have Wing Ng as mathematics department secretary. To all the conventional skills of a fine secretary she brings Excel and statistical expertise which she generously put to work on our project.

We appreciate the work of the following reviewers, who caught more than a few errors and contributed so many useful suggestions: Jeff Mock, Diablo Valley College (CA), Richard Spinetto, University of Colorado (CO), S. Christian Albright, Indiana University (IN), James Guffey, Truman State University (MO), David Shellabarger, Lane Community College (OR), Robert Johnson, Monroe Community College (NY), Roger Champagne, Hudson Valley Community College (NY), Hariharan K. Iyer, Colorado State University (CO), and Sam Kasala, University of North Carolina at Wilmington (NC). Finally, we want to acknowledge the fine work done by the Duxbury people—Curt Hinrichs, Cynthia Mazow, Tim Wardell, Alex Kugushev (our former editor), and our TEXguru William Baxter.

Millianne Lehmann
lehmann@usfca.edu

Paul Zeitz
zeitz@usfca.edu

San Francisco, California
August, 1997

Getting Started

1

In This Chapter...

- The Excel Window
- Workbooks and worksheets
- Toolbars: Standard and Formatting
- Selecting cells, ranges, rows, columns, and worksheets
- Entering data
- Formatting
- Clearing and editing cells
- Entering and copying formulas
- Sums and percentages
- Examples of absolute and relative cell addressing
- Text boxes
- Saving and retrieving a workbook
- Printing a spreadsheet selection

The goal of this chapter is the mastery of an important set of fundamental Excel skills. Begin by opening an Excel workbook. The procedure for doing so varies from machine to machine and depends on whether your computer is attached to a network or is a stand-alone machine, so we can't provide documentation here that will apply to all. If you don't know how to open Excel, ask a more experienced computer user for help.

1.1 THE EXCEL WINDOW

When Excel has been successfully loaded, you will see an Excel window open on the screen. The exact appearance of the window will vary depending on the version of Excel you are using. The following illustration shows the window as it looks in Excel 7, the version of Excel run by Windows 95.

EXCEL 7

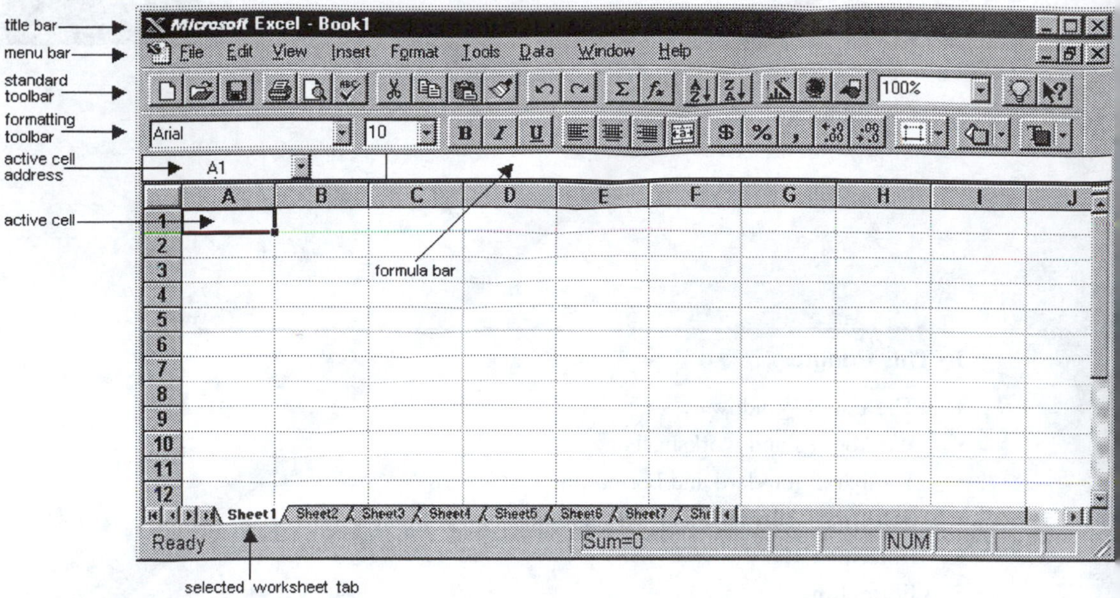

title bar
menu bar
standard toolbar
formatting toolbar
active cell address
active cell
formula bar
selected worksheet tab
Microsoft Excel - Book1
File Edit View Insert Format Tools Data Window Help
Ready
Sum=0
NUM

Important features of the Excel window are labeled. Familiarize yourself with them as they will be referred to often. The next illustration shows the Excel window as it appears on a Macintosh machine running Excel 5.

EXCEL 5

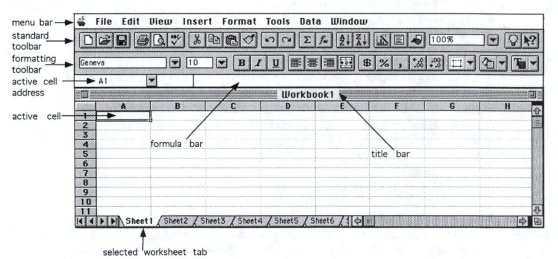

menu bar
standard toolbar
formatting toolbar
active cell address
active cell
formula bar
title bar
selected worksheet tab
File Edit View Insert Format Tools Data Window
Workbook1

The window for the latest (at this writing) version of Excel, Excel 97, is shown in the next illustration.

EXCEL 97

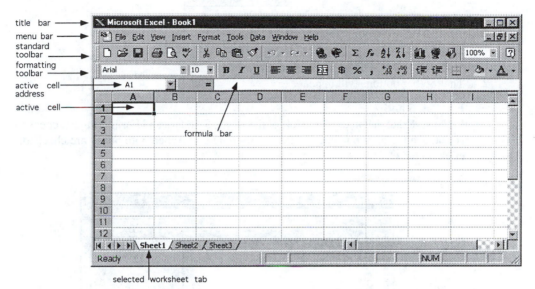

1.2 ADDING WINDOW FEATURES

Even if your display does not look *exactly* the same as any one of the pictures above, it should contain all of the labeled features. If not, it is usually a simple matter to add the missing ones.

What to Do If the Toolbars Are Missing

A toolbar is a line of buttons designed to automate especially useful Excel features. In this chapter you will use two toolbars. The first is the Standard Toolbar, which contains, among other options, buttons for ordinary computer tasks such as opening a file, printing a file, closing a file, and saving a file. The second is the Formatting Toolbar whose buttons facilitate, as its title suggests, a variety of formatting tasks such as control of typeface, alignment and borders, the number of decimal places shown in numeric data, the manner in which monetary amounts are displayed, and so forth.

If these toolbars are not on your screen, you must ask Excel to place them in view. To do this follow the steps below.

1. Pull down the View menu and select Toolbars... .

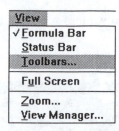

2. The toolbar options will appear.

    EXCEL 5 and 7
    Make sure that the boxes to the left of the words Standard and Formatting are checked.
    If not, use the mouse to click in the appropriate box. When both boxes are checked,
    click on the OK button.

EXCEL 97
The toolbar selections are made in a side menu as shown next.

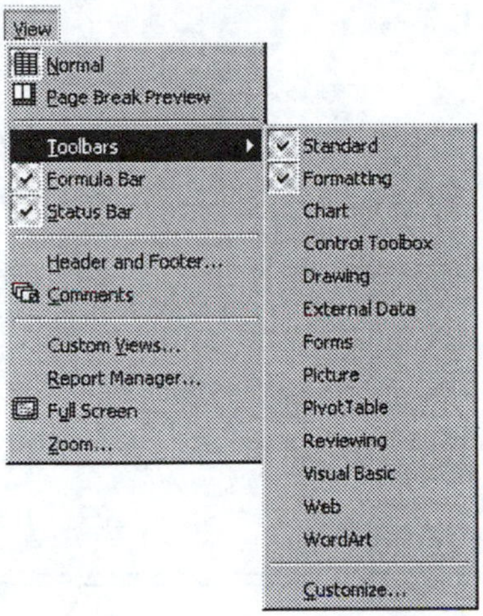

3. The checked toolbars will appear in your Excel window.

You can reposition a toolbar: All you have to do is click the toolbar anywhere between buttons, hold down the mouse button, and drag to the new location.

What to Do If the Workbook Title Is Missing

If you cannot find a workbook title such as "Book1" or "Workbook1" anywhere on your window, it means that your window does not yet contain an Excel workbook. If this is the case, you will have to open a new workbook. To do so just click on the New Workbook button shown below.

It is at the left end of the Standard Toolbar.

What to Do If the Gridlines Are Missing

If the workbook is blank, showing no horizontal and vertical gridlines, you will have to activate this feature. Follow the steps below to add them.

1. Pull down the Tools menu. Select Options. The Options dialog box will appear.
2. Click on the tab labeled **View** to bring the View options forward as shown below.

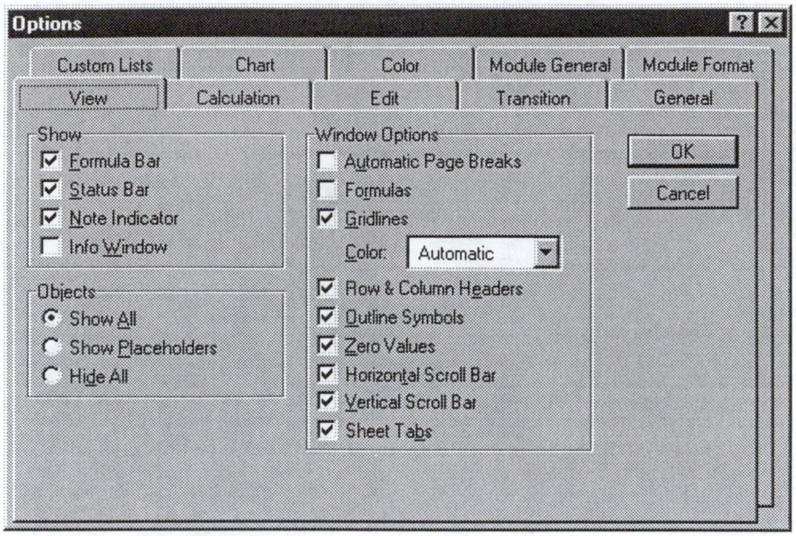

3.   Under **Window Options** click **Gridlines**.

4.   Click the OK button.

In Excel 97 the Gridlines option is at the lower left of View.

1.3 THE WORKBOOK

The heart of the Excel window is the *workbook*.

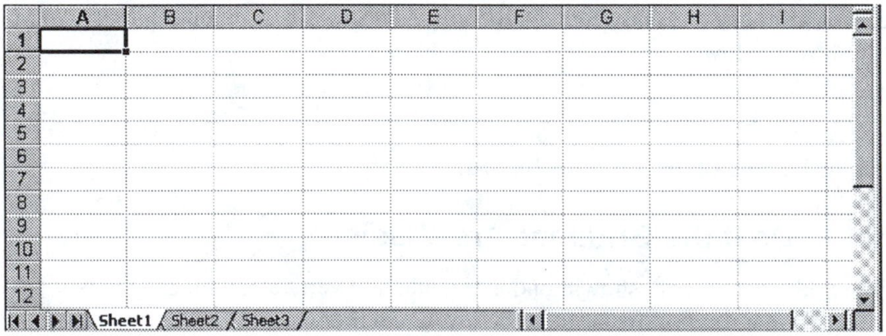

A workbook usually consists of several *worksheets*. For example, in the workbook shown above, **Sheet1** is the *selected* or *active worksheet*.

Selecting Worksheets

You can move from one worksheet to another in a workbook by clicking on the appropriate worksheet tab. Tabs are located on the bottom window border. At the left of the bottom

border are arrowheads, which provide an alternate method for moving among the
worksheets.

The Address of a Cell

A worksheet is divided into an array of *cells* created by the grid of horizontal and vertical
lines. Each cell has an address that consists of its column letter followed by its row number.
For example, the cell in the fourth column and the fifth row of the worksheet is referred to
as cell D5.

How to Select a Cell

To select a cell, move the cursor to the inside of the cell and click. In the worksheet below
the cell D5 has been selected.

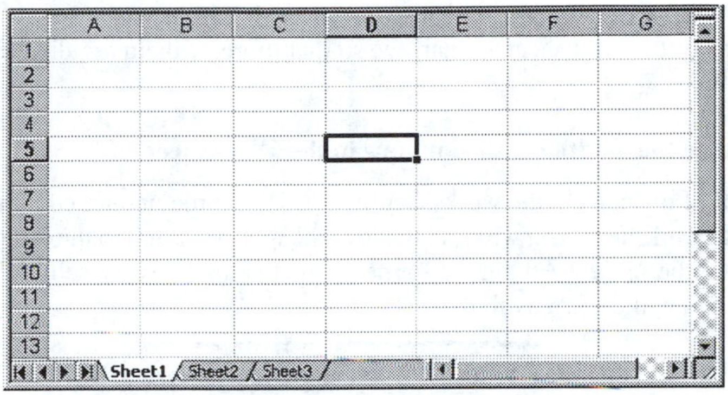

D5 is referred to as the *active cell*, meaning, among other things, that any typed characters
will be written in this cell.

How to Select a Range of Cells

The notation B2:D5 indicates the range of cells from B2 on the upper left to D5 on the
lower right. Follow the instructions below to select this range of cells.

1. Move the cursor to cell B2.
2. Hold down the mouse button and drag diagonally to cell D5.
3. Release the mouse and the selected range will be highlighted as in the next illustration.

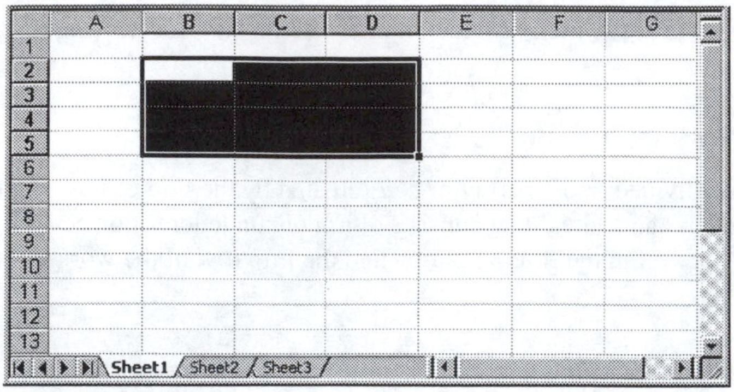

Adjusting Column Width

Shortly, you will type some data into this worksheet. It will be easier to see all of the data if the columns of the worksheet are narrow, so that more of them are displayed in your Excel window.

How to Change the Width of All Columns in the Worksheet

1. Select all of the cells in the worksheet by clicking on the unmarked button at the top left of the worksheet. You will find it above the number 1 and to the left of the letter A. It is called the "Select All button." When you click on it, every cell in the worksheet (except A1) will darken as shown below.

2. Pull down the Format menu, slide the cursor to the right at Columns and release on Width. The Width dialog box will appear. Type the number 5 into the Column Width box and click the OK button.

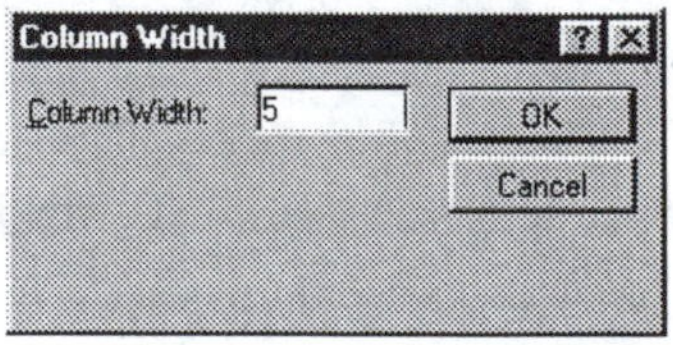

3. Your worksheet should now look like the one pictured below.

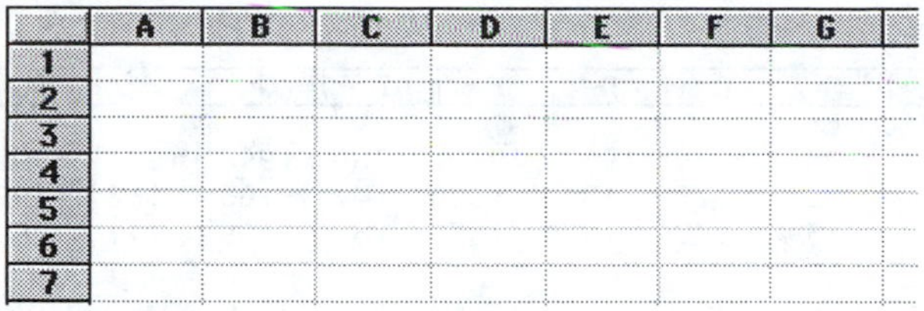

How to Change the Width of a Single Column

As practice, widen column A.

1. Position the cursor at the line dividing column A from column B. When the cursor is in the correct position, it will become a black cross with arrow heads on the horizontal bar of the cross as shown below.

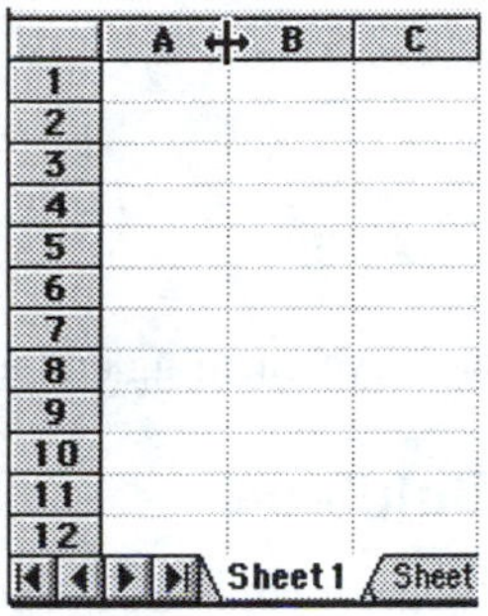

2. Click the mouse and drag to the right until the column looks wider. Release the mouse.

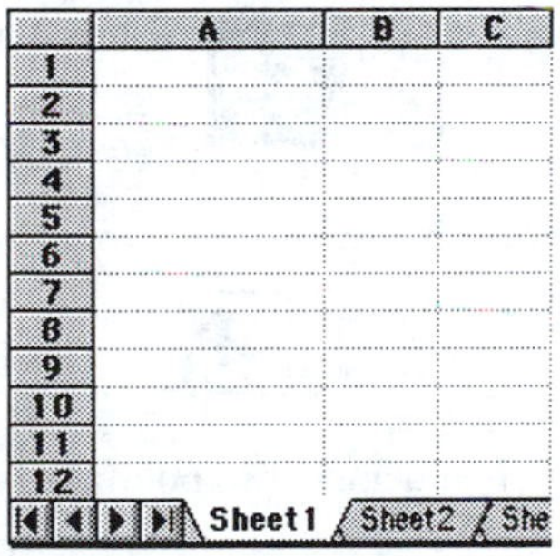

1.4 DATA ENTRY

In this section you will enter the table of data shown below. It gives the enrollment figures for women and men in the science college of a midsize American university. If you are working with a lab partner, one of you can dictate the data to the other. This will speed up the data-entry process.

	A	B	C	D	E	F	G	H	I	J	K	L
1	College of Science											
2	Year	1985	1986	1987	1988	1989	1990	1991	1992	1993	1994	1995
3	Women	221	175	139	130	147	163	168	192	244	264	280
4	Men	339	273	197	165	156	169	171	192	200	210	214

1. Select cell A1: Move the point of the cursor inside the cell and click.
2. Type "College of Science." (If you make an error, press the DELETE or backspace key until the error is erased and retype.) Notice that what you type appears both in the active cell and in the formula bar.
3. Press either the RETURN key or the ENTER key to complete the entry.
4. Select cell A2.
5. Type "Year" and then press the RETURN key or the ENTER key.
6. Continue until the table has been entered.

1.5 FORMATTING

Spruce up the table a bit to make it eye-catching and a little easier to read.

Change Column A to Boldface

1. Click on the letter A at the head of column A to select the entire column.

	A	B	C
1	College of Science		
2	Year	1985	1986
3	Women	221	175
4	Men	339	273

2. Click the Bold button shown below.

Now the entries in column A should appear in boldface type.

	A	B	C
1	**College of Science**		
2	**Year**	1985	1986
3	**Women**	221	175
4	**Men**	339	273

Widen column A a little so that the "n" in "Women" is visible.

Write the Title in Italic

Select cell A1 by clicking on it and then click on the Italic button shown next.

"College of Science" should now be written in italic.

Center Row 2 and Give It a Border

1. Select the row of years by clicking on the row number "2."

	A	B	C	D	E	F	G	H	I	J	K	L
1	*College of Science*											
2	Year	1985	1986	1987	1988	1989	1990	1991	1992	1993	1994	1995
3	Women	221	175	139	130	147	163	168	192	244	264	280
4	Men	339	273	197	165	156	169	171	192	200	210	214
5	Total	560	448	336	295	303	332	339	384	444	474	494

2. Press the Bold button twice (Two clicks are needed because the row is originally a mix of both bold- and normal-face entries.) and then the Center button which is shown in the following picture.

3. With row 2 still selected click the drop-down arrow on the Border button and select the heavy lower border as illustrated in the next picture.

    EXCEL 5 and 7

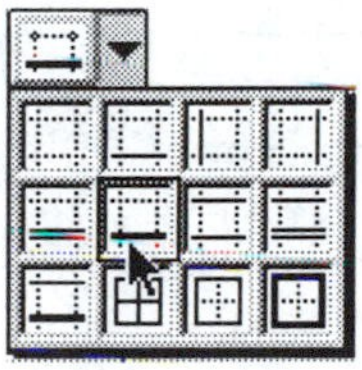

    In Excel 97 the Border button menu looks a little different; however, just as above select the second border option from the left in the second row.

    EXCEL 97

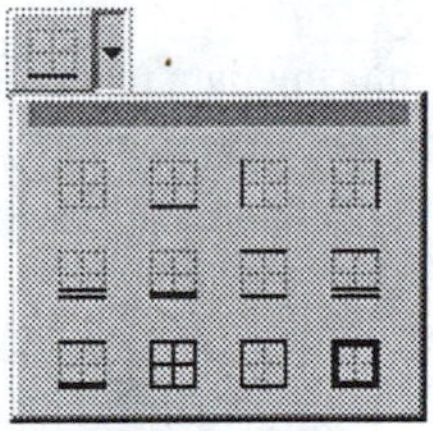

    Your table should now look something like the picture below.

	A	B	C	D	E	F	G	H	I	J	K	L
1	*College of Science*											
2	Year	1985	1986	1987	1988	1989	1990	1991	1992	1993	1994	1995
3	Women	221	175	139	130	147	163	168	192	244	264	280
4	Men	339	273	197	165	156	169	171	192	200	210	214

Center the Row Headings

1. Select the range A3:A4 by clicking in cell A3 and dragging to cell A4 so that both cells are selected as shown next.

	A	B	C
1	*College of Science*		
2	Year	1985	1986
3	Women	221	175
4	Men ⊕	339	273

2. Press the Center button.

More Borders

Select the range A2:A4 in this fashion: click in cell A2 and drag to cell A4. Click the right-hand Border button from the drop-down Border menu. The formatted table should look like the next illustration.

	A	B	C	D	E	F	G	H	I	J	K	L
1	*College of Science*											
2	Year	1985	1986	1987	1988	1989	1990	1991	1992	1993	1994	1995
3	Women	221	175	139	130	147	163	168	192	244	264	280
4	Men	339	273	197	165	156	169	171	192	200	210	214

1.6 MAKING CORRECTIONS

You are bound to make errors when entering data. Here are a few tips for correcting the inevitable goofs.

1. If you make an error while entering data in a cell, you can cancel the entry by clicking the Cancel box. (The Cancel box is visible only while data is being entered.) It is found in the group of buttons to the left of the formula bar as shown next.

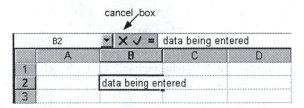

2. The contents of a cell can be overwritten. Select the cell whose contents you wish to change and then type in the new material. Press the ENTER key.

3. The contents of a cell can be edited. Select the cell to be edited so that its contents are displayed in the formula bar. Click in the formula bar display and make the changes you wish. When editing is complete, press the ENTER key.

4. If you wish to completely erase the contents of a cell or range of cells, select the cell or range and then select Clear Contents from the Edit menu.

5. You can remove formatting but keep the contents of cells: select the range of cells and then pull down the Edit menu. Select Clear and then Format.

6. When you select Clear and then All from the Edit menu, Excel will remove the contents as well as the formatting from selected cells.

1.7 SOME SPREADSHEET CALCULATIONS

Run your eye over the enrollment data and, as you do so, imagine yourself a student or faculty member interested in promoting careers for women in science. You notice a severe decline in female enrollment during the second half of the 1980s. This decline was not reversed until 1993. However, comparing male to female enrollments, it is easy to see similar changes in the data for men. So, in evaluating the status of female enrollments over the 11-year period it is important to look at the changes in the *proportion* of women in the science college student body. Proportion constitutes a more *valid* measure of female status relative to male than the raw numbers alone.

Finding Sums

Calculate the total number of students in the college for each of the 11 years.

1. Click in cell A5 to select it. Type the label "Total" and change it to boldface.

2. Click in cell B5. Type the formula =B3+B4 (don't forget the equal sign) and press the ENTER key.

B5		▼		**=** =(B3+B4)
	A	B	C	D	E
1	*College of Science*				
2	Year	1985	1986	1987	1988
3	Women	221	175	139	130
4	Men	339	273	197	165
5	Total	560	448	336	295
6					

3. Copy the formula in cell B5 to the range B5:L5 as follows: Select cell B5 and move the cursor to the cell's lower right corner. When it becomes a black cross, click and drag across to cell L5 and release the mouse. Excel will automatically calculate the total enrollment for each year.

B5			=	=(B3+B4)							
	A	B	C	D	E	F	G	H	I	J	K	L
1	*College of Science*											
2	Year	1985	1986	1987	1988	1989	1990	1991	1992	1993	1994	1995
3	Women	221	175	139	130	147	163	168	192	244	264	280
4	Men	339	273	197	165	156	169	171	192	200	210	214
5	Total	560	448	336	295	303	332	339	384	444	474	494
6												

If Excel doesn't calculate these sums for you, repeat the previous step making sure that the black cross is visible as you drag. Still doesn't work? Try holding down the SHIFT key and pressing the F9 key. No luck? We give up—ask for help.

Finding Percentages

Calculate the percentage of women enrolled each year.

1. Click cell A7. Type "% Women" and press the ENTER key. Press the Bold button.
2. Click cell A8. Type "% Men" and press the ENTER key. Press the Bold button.
3. Click in cell B7. Type the formula =B3/B5 and press the ENTER key.

B7			=	=B3/B5
	A	B	C	D	E
1	*College of Science*				
2	Year	1985	1986	1987	1988
3	Women	221	175	139	130
4	Men	339	273	197	165
5	Total	560	448	336	295
6					
7	%Women	0.39			
8	%Men				

4. Change the fraction to a percentage by clicking on cell B7 and pressing the Percent button. It is shown below.

%

Check your worksheet against the one displayed next. The entry in cell B7 should be 39%.

B7			=	=B3/B5
	A	B	C	D	E
1	*College of Science*				
2	Year	1985	1986	1987	1988
3	Women	221	175	139	130
4	Men	339	273	197	165
5	Total	560	448	336	295
6					
7	%Women	39%			
8	%Men				

5. Copy the formula in B7 to range B7:L7. Select cell B7 and
    the lower right corner of the selected cell. When the curso
    click and drag to cell L7. Excel will automatically calculate
    each year's enrollment and express it as a percentage of tota

B7		▼		=	=B3/B5			
	A	B	C	D	E	F	G	H	I
1	College of Science								
2	Year	1985	1986	1987	1988	1989	1990	1991	1992
3	Women	221	175	139	130	147	163	168	192
4	Men	339	273	197	165	156	169	171	192
5	Total	560	448	336	295	303	332	339	384
6									
7	%Women	39%	39%	41%	44%	49%	49%	50%	50%	55
8	%Men								

6. Complete your table by filling in the %Men row. Your final table should look
    something like the illustration below.

	A	B	C	D	E	F	G	H	I	J	K	L
1	College of Science											
2	Year	1985	1986	1987	1988	1989	1990	1991	1992	1993	1994	1995
3	Women	221	175	139	130	147	163	168	192	244	264	280
4	Men	339	273	197	165	156	169	171	192	200	210	214
5	Total	560	448	336	295	303	332	339	384	444	474	494
6												
7	%Women	39%	39%	41%	44%	49%	49%	50%	50%	55%	56%	57%
8	%Men	61%	61%	59%	56%	51%	51%	50%	50%	45%	44%	43%

Reflections on the Data

These percentages provide an interesting insight into the data: although the *number* of
women fell dramatically in the early years and rose in later years, the *percentage* of
women enrolled *never* fell. On the contrary, the proportion of women rose steadily over
the entire 11-year period. The alarming fall in female enrollment during the late 1980s was
a reflection of the overall precipitous decline in science students.

1.8 OPENING A NEW WORKSHEET

We have finished with the enrollment data worksheet for the time being. In the next section
we will be investigating a new set of data and will need a fresh worksheet to hold it.

Using Sheet Tabs

Look down at the bottom of your Excel worksheet at the line of sheet tabs. The tab for
the active sheet is displayed in boldface. In the illustration below, **Sheet1** contains the
enrollment data.

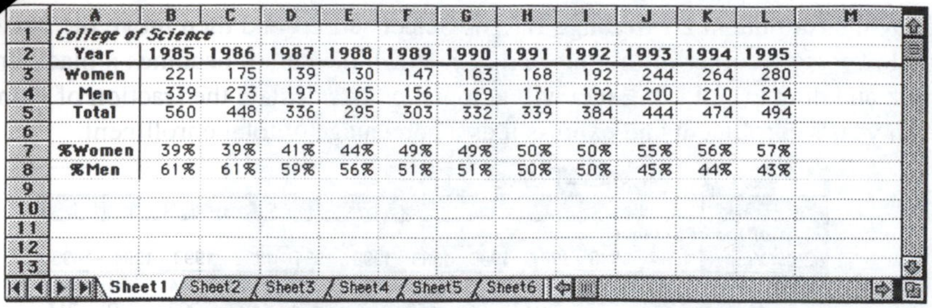

	A	B	C	D	E	F	G	H	I	J	K	L	M
1	College of Science												
2	Year	1985	1986	1987	1988	1989	1990	1991	1992	1993	1994	1995	
3	Women	221	175	139	130	147	163	168	192	244	264	280	
4	Men	339	273	197	165	156	169	171	192	200	210	214	
5	Total	560	448	336	295	303	332	339	384	444	474	494	
6													
7	%Women	39%	39%	41%	44%	49%	49%	50%	50%	55%	56%	57%	
8	%Men	61%	61%	59%	56%	51%	51%	50%	50%	45%	44%	43%	
9													
10													
11													
12													
13													

Sheet1 / Sheet2 / Sheet3 / Sheet4 / Sheet5 / Sheet6

Click on the **Sheet2** tab to open a new worksheet.

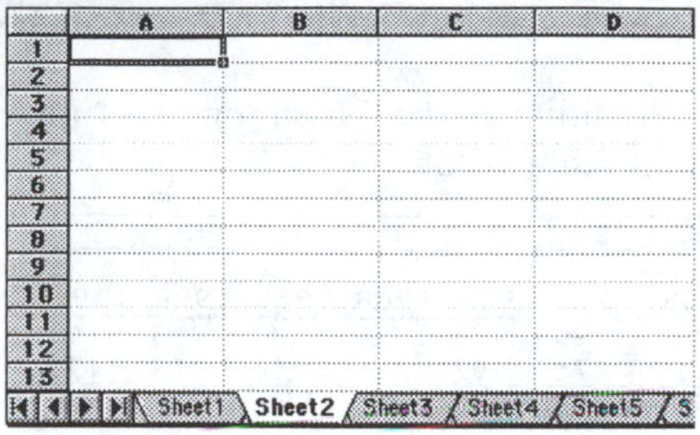

	A	B	C	D
1				
2				
3				
4				
5				
6				
7				
8				
9				
10				
11				
12				
13				

Sheet1 \ Sheet2 / Sheet3 / Sheet4 / Sheet5 / S

Return to **Sheet1** by clicking on its tab.

Using the Insert Menu

If only one tab is visible at the bottom of your workbook, you can still open a new worksheet. All you have to do is pull down the Insert menu and select New Worksheet.

In the next chapter, we will have a good deal more to say about worksheet management. For the time being, just think of the various worksheets as separate pages in your Excel workbook.

1.9 AN EXAMPLE OF ABSOLUTE AND RELATIVE CELL ADDRESSING

Type the data shown below into a new worksheet. (Practice your formatting skills—make your table look like the illustration.) The table gives the number of years of computer experience reported by the entering class of a California university. Notice that only 39 students reported that they had no computer experience at all, whereas 240 said they had used a computer for more than 5 years.

	A	B	C
1	*Computer Experience of New Students*		
2	Experience	Number of	
3	in Years	Students	
4	0	39	
5	up to 1	130	
6	1 to 3	309	
7	3 to 5	205	
8	over 5	240	
9	no response	5	
10	Total		
11			

A Digression—The AutoSum Button

Use the AutoSum button to calculate the total number of new students.

1. Select cell B10 and click the AutoSum button shown below.

It is a Standard Toolbar button.

2. Excel will automatically place the command =SUM(B4:B9) into cell B10.

	A	B	C
1	*Computer Experience of New Students*		
2	Experience	Number of	
3	in Years	Students	
4	0	39	
5	up to 1	130	
6	1 to 3	309	
7	3 to 5	205	
8	over 5	240	
9	no response	5	
10	Total	=SUM(B4:B9)	
11			

A dashed border appears around the range of cells in the SUM command. Check to make sure that these are the numbers whose total is required.

3. The range is correct, so press the ENTER key.

	A	B	C
1	*Computer Experience of New Students*		
2	Experience	Number of	
3	in Years	Students	
4	0	39	
5	up to 1	130	
6	1 to 3	309	
7	3 to 5	205	
8	over 5	240	
9	no response	5	
10	Total	928	
11			

We now know that there were 928 new students in all. The next task is to calculate the percentage of new students in each of the six computer usage groups and to do so efficiently by employing both *absolute* and *relative* cell addresses in the formula we use. Simply described, an absolute cell address is one that does not change when the formula containing it is copied from one position in a worksheet to another. A relative cell address, on the other hand, will change when a formula is copied. Absolute cell addresses are indicated by $ signs. For example, $A$1 will not change when copied, whereas A1 will. The address $A1 is *mixed*. In this case, the column address, A, will not change while the row address, 1, will change.

As a first example of the use of these address types, study the calculations below. There is much more to learn about cell addressing, and we will return to this topic again in the exercises and in later chapters. This is a beginning.

1. Type the words "Percent of" in cell C2 and "New Students" in cell C3. Add a lower border to cells C3 and C9.

2. Now, let's stop a minute and think about the calculations we want Excel to make in the Percent column. The formulas needed are shown below in range C4:C10.

	A	B	C
1	*Computer Experience of New Students*		
2	**Experience**	**Number of**	**Percent of**
3	**in Years**	**Students**	**New Students**
4	0	39	=B4/B10
5	up to 1	130	=B5/B10
6	1 to 3	309	=B6/B10
7	3 to 5	205	=B7/B10
8	over 5	240	=B8/B10
9	no response	5	=B9/B10
10	Total	928	=B10/B10

Notice that each formula contains a fraction that gives the proportion of the total number of new students who have the indicated level of computer experience. As you move from one row to the next, the row address in the denominator does not change. It remains fixed at the number 10. The row addresses in the numerators behave differently. They change so that the number of students in each successive level of computer experience becomes the numerator in turn.

Since there are only six formulas involved here, one might be tempted to type in all six and be done with it. However, efficiency is our goal, so we will type only the formula for cell C4, but craft it so that when it is copied down the column it produces the six required formulas automatically. The formula =B4/B$10 will do the job.

3. Enter the formula =B4/B$10 into cell C4 and press the ENTER key. Type the formula carefully, placing the $ sign in the denominator of the fraction only. The row address, 10, must be absolute.

4. Select cell C4 again and press the Percent button on the toolbar. The result is shown below.

C4		=	=B4/B$10	
	A	B	C	D
1	*Computer Experience of New Students*			
2	**Experience**	**Number of**	**Percent of**	
3	**in Years**	**Students**	**New Students**	
4	**0**	39	4%	
5	**up to 1**	130		
6	**1 to 3**	309		
7	**3 to 5**	205		
8	**over 5**	240		
9	**no response**	5		
10	**Total**	928		

The calculation tells us that 4% of all new students reported no computer experience.

5. Copy this formula to each cell in the column range C4:C10. Select cell C4. Move the cursor to the lower right corner of cell C4. When it becomes a black cross, double-click to copy the formula down the column. (Copying down can be automated in this fashion. Alternately, you can use a black-cross drag to copy the formula from C4 to C10.) In either case, Excel will automatically calculate the percentage of new students in each computer-use category.

C7		=	=B7/B$10	
	A	B	C	D
1	*Computer Experience of New Students*			
2	**Experience**	**Number of**	**Percent of**	
3	**in Years**	**Students**	**New Students**	
4	**0**	39	4%	
5	**up to 1**	130	14%	
6	**1 to 3**	309	33%	
7	**3 to 5**	205	22%	
8	**over 5**	240	26%	
9	**no response**	5	1%	
10	**Total**	928	100%	

1.10 THE TEXT BOX

A *text box* is a special area created in a worksheet where comments and descriptions can be inserted. It functions like a mini word processor embedded in the worksheet. For practice with text boxes, embed one in the "computer experience" worksheet discussed in the previous section.

How to Embed a Text Box

1. Click on the Drawing button. It is a Standard Toolbar button and will resemble one of the buttons shown below depending on the version of Excel you are using.

2. The Drawing Toolbar will open on the worksheet. Click the Text Box button, which will look like one of the illustrations below, again depending on the version of Excel you are using.

3. Click and drag diagonally in your worksheet and a rectangular box will appear. This rectangle is the text box. Release the mouse when the box is the size and shape you want.

4. Close the Drawing Toolbar by clicking on its Close box—the cross at the upper right corner of the toolbar.

5. Click inside the text box and begin to type.

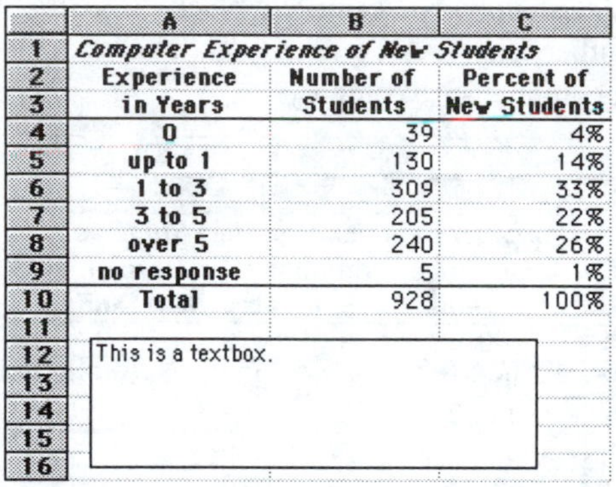

	A	B	C
1	*Computer Experience of New Students*		
2	**Experience**	**Number of**	**Percent of**
3	**in Years**	**Students**	**New Students**
4	0	39	4%
5	up to 1	130	14%
6	1 to 3	309	33%
7	3 to 5	205	22%
8	over 5	240	26%
9	no response	5	1%
10	Total	928	100%
11			
12	This is a textbox.		
13			
14			
15			
16			

How to Resize a Text Box

The size of your text box can be changed. Place the point of the cursor on the border of the text box and click. The border will thicken (see the next illustration) and rectangular handles will be displayed at the corners of the box and in the middle of each side.

11	
12	This is a textbox.
13	
14	
15	
16	

Click and drag on one of these handles to alter the dimensions of the box.

How to Reposition a Text Box

Place the point of the cursor anywhere on the border of the text box. Press the mouse and drag the box to the place where you want it on the worksheet. Release the mouse.

How to Remove a Text Box

You can remove a text box from your worksheet by first clicking on its border and then pressing the DELETE key.

1.11 SAVING, PRINTING, AND RETRIEVING AN EXCEL WORKBOOK

As you type information into a workbook and perform spreadsheet calculations, it is important that you pause periodically and save your work. How this is done depends on the nature of the system you are using and on your personal preferences. For example, if you are working on your own computer, you can save your workbook on the machine's internal hard drive. However, if you want the workbook to be portable because, for example, you are going to complete your work later in a computer lab, then the document should be saved to a diskette. If your Excel work is done on a computer attached to a network, then you might have the option of saving copies of your Excel workbooks in a user file or directory. In every case the general procedure for the initial saving of a file is the same—pull down the File menu and release on Save As... . When the Save As dialog box appears, name your workbook and select a destination for the document and click on Save. If you then continue working and, a little later, decide to save your workbook again, simply pull down the File menu and select Save (not Save As...).

A saved workbook is retrieved by pulling down the File menu and releasing on Open. When the Open dialog box appears, you must select the options that tell Excel the location of the document you are trying to reopen.

Printing a Selected Range of Cells

1. Highlight the portion of the worksheet that you want to print.

	A	B	C	D
1	*Computer Experience of New Students*			
2	Experience	Number of	Percent of	
3	in Years	Students	New Students	
4	0	39	4%	
5	up to 1	130	14%	
6	1 to 3	309	33%	
7	3 to 5	205	22%	
8	over 5	240	26%	
9	no response	5	1%	
10	Total	928	100%	
11				
12	This is a textbox.			
13				
14				
15				
16				
17				
18				

2. Pull down the File menu and select Print.
3. When the Print dialog box opens, click on **Selection**. The exact appearance of the dialog box will vary from machine to machine. Two examples follow and, in both cases, **Selection** has been checked.

EXCEL 5

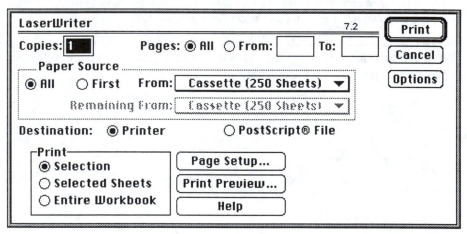

EXCEL 7

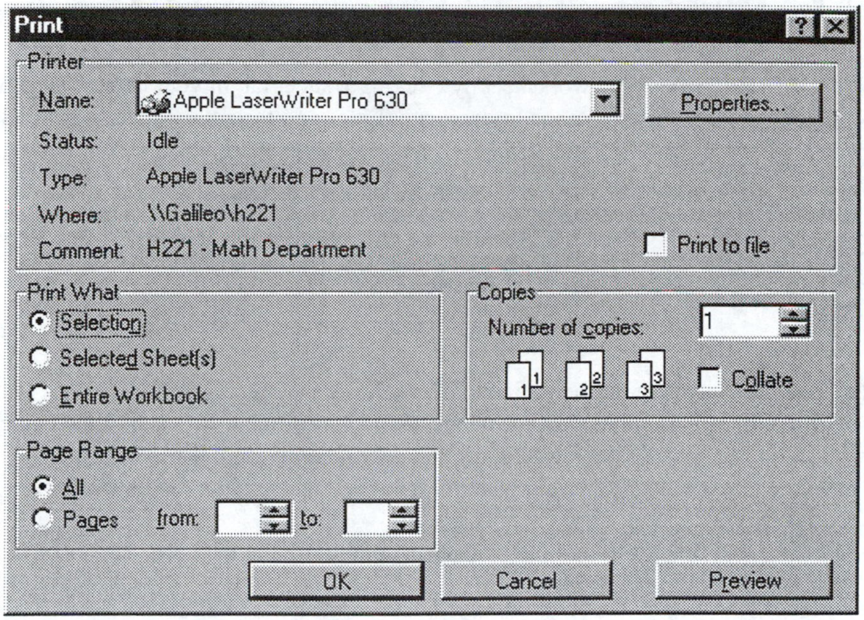

EXCEL 97
In Excel 97 the Selection option is in the lower-left corner of the Print dialog box.

4.   Make sure that the number of copies is set at 1. Click the OK or the Print button.

EXERCISES

Before printing an exercise, be sure you type your name somewhere in the print region. You may have classmates using the same printer, and you need a way to distinguish your work from theirs.

Exercise 1.1 Follow the instructions in Section 1.9 to create the "computer experience worksheet" that appears at the end of that section. Select and print just the final table.

Exercise 1.2 If you haven't already done so, enter the College of Science enrollment data from Section 1.4 in a new worksheet.
 (a) Work through the calculation of the percentages of male and female enrollment as described in Section 1.7.
 (b) Use the same data to calculate the ratio of female enrollment to male enrollment in percentage terms for each year.
 (c) Open a text box and comment on any evidence of gender discrimination you see in this data. What additional information would be useful in evaluating a claim of gender

discrimination in the College of Science enrollments?
(d) Select just that portion of the worksheet containing the data, the percentages, and the text box. Print your selection.

Exercise 1.3 Open a new worksheet by clicking a sheet tab in your current workbook or by selecting Worksheet from the Insert menu. Enter and format the data shown below.

	A	B	C
1	Enrollment by College		
2	Arts	800	
3	Business	488	
4	Nursing	278	
5	Science	368	
6	Undecided	112	
7	Total		

(a) Use the AutoSum button to find total enrollment.
(b) In column C calculate the percentage of total enrollment for each of the colleges by typing the appropriate formula into cell C2 and copying it down to cell C7.
(c) Open a text box and explain what forms of cell addressing you used in your formula and why.
(d) Select just that portion of the worksheet containing the data, the percentages, and the text box. Print your selection.

Exercise 1.4 Open a new worksheet by clicking a sheet tab in your current workbook or by selecting Worksheet from the Insert menu. Enter and format the data shown below. In the early 1970s the State of Oklahoma used the data in this table as part of its defense of an Oklahoma law that prohibited the sale of 3.2% beer to males under the age of 21 but permitted its sale to females aged 18 to 21.[1]

	A	B	C	D	E	F	G
1							
2	Arrests by Age and Sex in Oklahoma, September to December 1973						
3			MALES			FEMALES	
4		18-21	Over 21	Total	18-21	Over 21	Total
5	Driving under influence	427	4973	5400	24	475	499
6	Drunkenness	966	13,747	14,713	102	1176	1278

As the first step in investigating this data, calculate the totals in each of the data columns.
(a) Click in cell A7 and type the word "Total."
(b) Click in cell B7 and then click the AutoSum button. Press the ENTER key.
(c) Move the cursor to the lower right corner of cell B7. When it becomes a black cross, click and drag to cell G7. The various totals for male and female arrests will appear in row 7 as shown below.

---

[1] Adapted from *Seeing Through Statistics* by Jessica Utts, Duxbury Press, 1996, Case Study 6.3

	A	B	C	D	E	F	G
1							
2	Alcohol Related Arrests by Age and Sex in Oklahoma, September to December 1973						
3				**MALES**		**FEMALES**	
4		**18-21**	**Over 21**	**Total**	**18-21**	**Over 21**	**Total**
5	Driving under influence	427	4973	5400	24	475	499
6	Drunkenness	966	13,747	14,713	102	1176	1278
7	Total	1393	18,720	20,113	126	1651	1777

(d) Complete the formatting: use the Border button to place a bottom border on cells A6:G6 and a right-hand border on cell D7.

(e) Looking only at these totals, what argument might Oklahoma have used to explain its law forbidding the sale of 3.2 beer to men aged 18 to 21, but not to women? Place your answer in a text box and print it along with the table. (The next exercise uses this table so keep your worksheet open.)

Exercise 1.5 In this exercise you will use the data from the previous problem to produce a table of the percentages of men and of women in each age category. As you will see, these percentages give a clearer picture of the incidence of male and female alcohol-related arrests. Follow the steps outlined below.

(a) Prepare the titles for the new table. Highlight the range A2:G4—click in cell A2 and drag to cell G4. Pull down the Edit menu and select Copy. Click in cell A9 and then pull down the Edit menu and select Paste. Now copy the contents of A5:A7 to A12:A14 in a similar manner. The result is shown next.

	A	B	C	D	E	F	G
1							
2	Alcohol Related Arrests by Age and Sex in Oklahoma, September to December 1973						
3				**MALES**		**FEMALES**	
4		**18-21**	**Over 21**	**Total**	**18-21**	**Over 21**	**Total**
5	Driving under influence	427	4973	5400	24	475	499
6	Drunkenness	966	13,747	14,713	102	1176	1278
7	Total	1393	18,720	20,113	126	1651	1777
8							
9	Alcohol Related Arrests by Age and Sex in Oklahoma, September to December 1973						
10				**MALES**		**FEMALES**	
11		**18-21**	**Over 21**	**Total**	**18-21**	**Over 21**	**Total**
12	Driving under influence						
13	Drunkenness						
14	Total						

(b) Calculate the number of men arrested, aged 18 to 21, as a percentage of all male arrests. Enter the formula =B5/$D$7 into cell B12. Press the Percent button. The result should be 2%. What does this percentage mean?

(c) Select cell B12 and place the cursor at its lower right corner. When the cursor changes to a black cross, click and drag to cell D12. With the range B12:D12 selected, place the cursor at the lower right corner of D12. When the cursor changes to a black cross, click and drag to cell D14. Your worksheet should look like the next illustration.

	A	B	C	D	E	F	G
1							
2	Alcohol Related Arrests by Age and Sex in Oklahoma, September to December 1973						
3			MALES			FEMALES	
4		18-21	Over 21	Total	18-21	Over 21	Total
5	Driving under influence	427	4973	5400	24	475	499
6	Drunkenness	966	13,747	14,713	102	1176	1278
7	Total	1393	18720	20113	126	1651	1777
8							
9	Percent of Alcohol Related Arrests by Age and Sex in Oklahoma, September to December 1973						
10			MALES			FEMALES	
11		18-21	Over 21	Total	18-21	Over 21	Total
12	Driving under influence	2%	25%	27%			
13	Drunkenness	5%	68%	73%			
14	Total	7%	93%	100%			

(d) Now make the same percentage calculations for the female portion of the table. Print the table. (The next exercise uses this table so keep your worksheet open.)

Exercise 1.6 This exercise refers to the table produced in the previous problem. Recall that in the early 1970s the State of Oklahoma used the data in this table as part of its defense of an Oklahoma law that prohibited the sale of 3.2% beer to males under the age of 21 but permitted its sale to females aged 18 to 21. The State claimed that the data showed that young men were more likely to be involved in alcohol-related arrests than were young women. The law was challenged by a young man who claimed discrimination and the case ended up in the United States Supreme Court. If you had been an attorney for the young man, how would you have used the percentage table to attack the law? What additional data might have been useful? Write your answer in a text box, select it and print it along with any tables you refer to in your discussion. Everything should fit neatly on one page.

Exercise 1.7 Excel has a built-in prompt that can help you to keep track of the toolbar buttons and the tasks which they are designed to expedite. Activate this feature by bringing the Status Bar into view: pull down the View menu and make sure that the line Status Bar is checked. If not, select it. Once Status Bar is checked, move the cursor to any toolbar button and keep it there for a half-second or so. Excel will display the name of the button and, at the bottom of the screen, the Status Bar will contain a brief description of what the button does. Experiment with each of the buttons listed below and write a sentence or two in a text box explaining each button's effect. (In Excel 97, the Status Bar does *not* contain a description of the button, a dubious "improvement.")
 (a) Zoom Control
 (b) Currency Style
 (c) Undo
 (d) Comma Style
 (e) Font Size

Exercise 1.8 Type the following four numbers into the first four rows of any column of a new worksheet: your age, your social security number, your zip code, and your weight to the nearest pound. Select the four numbers and click on one of the two Sort buttons and then the other. In another column type the following four words: your major, your last name, your native city, and your mother's maiden name. Try the two Sort buttons on this list. Explain the effects of these buttons in a text box. Select the four numbers, the four words, and the text box, and print the selection.

WHAT TO HAND IN: For each exercise assigned, print just that portion of the worksheet asked for. You should have no more than one page per exercise.

Numerical Measures
of Data

2

In This Chapter…

- Opening a **datach** workbook from the **STAT** diskette
- Opening, naming, splitting, inserting, and deleting worksheets
- Selecting data from large worksheets
- Read-Only files
- Managing workbooks and worksheets
- COUNT, COUNTIF, and IF
- Functions in Excel
- MEAN, MEDIAN, and MODE
- Function Wizard and Paste Function
- Pitfalls in collecting survey data
- Five-Number Summaries
- MAX, MIN, QUARTILE, and PERCENTILE

In this chapter you will have the opportunity to analyze some results from a survey of several hundred university students. The survey consisted of 20 questions covering a variety of subjects. Each student was asked about his or her college and high school grade point average and for vital statistics including height, weight, gender, and age. Questions about smoking habits, exercise, pleasure reading, and driving skills were also asked. In addition to these common questions asked of everyone, there were also a few questions posed in two different forms so that half the students responded to one version of the item and half the students to another. These questions were included in order to gather information about the process of surveying itself.

Before going on install STAT, the diskette that accompanies this book, on your computer. See Appendix B for instructions.

Begin by opening the workbook **datach2.xls**. Follow the sequence of steps outlined next.

2.1 HOW TO RETRIEVE A WORKBOOK FROM STAT

1. Follow the instructions in Appendix B to install **STAT** on your computer.
2. If Excel is not already running on your computer, open it.
3. Pull down Excel's File menu and select Open.
4. When the Open dialog box appears, select the internal hard drive (probably drive C).
5. Double-click on **STAT** to display its contents.
6. Finally, double-click on **datach2.xls** and the workbook will open on your screen.

2.2 HOW TO SURVIVE IN A BIG WORKSHEET

Make sure that the workbook **datach2.xls** is open and click on the **student data** worksheet tab so that the survey data shown below is displayed.

datach2.xls [Read-Only]							
	A	B	C	D	E	F	G	H
1	*Statistical Reasoning Class Survey*							
2	**Section 5, * means no response**							
3	College				More Likely?		Driving Skill	
4	Liberal Arts = 1	High		Height	HHTHTT=1	Gender	Better = 1	
5	Business=2, Nursing = 3	School	Age in	in	HHHHHH=2	male = 1	Worse = 3	Weight in
6	Science=4, Undecided=5	GPA	Years	Inches	Equally=3	fem = 2	Average = 2	Pounds
7	5	*	21	*	2	2	3	112
8	2	3.2	19	65	1	2	2	120
9	1	3.3	19	66	1	2	3	123
10	2	*	20	69	3	2	1	130
11	2	1.2	21	63	1	2	1	105
12	2	3.5	19	*	1	2	2	120
13	5	3.7	18	65	3	2	1	135
14	3	3.3	28	69	1	1	1	160
15	2	*	18	60	3	2	2	110
16	3	3	29	62	3	2	2	108
17	3	3.4	19	61	3	2	1	109

student data / order of questions-moon dia. / driving-gender

Unless you have a very large monitor, this worksheet is both too wide and too long to fit on the screen.

Maximizing the Window

Make sure that your window is *maximized*, so that both the right and bottom scroll bars are visible. This can be accomplished by clicking on the Maximize button, which, on a Windows machine, looks like the next picture.

On a Macintosh computer, look for the button shown below.

Once the scroll bars are visible, you can use them to scan through the data. The PAGE UP and PAGE DOWN keys are also helpful in moving about a large worksheet. However, by far the most useful and efficient method of large worksheet management is with a *split screen.*

Splitting the Screen

Using the mouse, move the cursor to the small strip at the top of the vertical scroll bar. When the cursor is in the top position it takes on the cross shape pictured below.

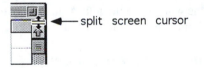
split screen cursor

Keeping the split screen cursor visible, press the mouse and drag down inside the scroll bar past a few rows. When you stop pressing the mouse button, the screen will divide into two sections. Each section, top and bottom, has its own scroll bar at the right-hand edge of the window. Scroll down the lower section to the bottom (the fastest way to do this is to drag the scroll box), so that you can see both the first and last rows of the data at the same time. Your worksheet should now look something like the next picture.

	A	B	C	D	E	F	G	H	I
1	Statistical Reasoning Class Survey								
2	Section 5, * means no response								
3	College						Driving Skill		Moons Diameter
4	Liberal Arts = 1	High		Height	More Likely? HHTHTT=1	Gender	Better = 1		Estimate
5	Business=2, Nursing = 3	School	Age in	in	HHHHHH=2	male = 1	Worse = 3	Weight in	<2000=2
6	Science=4, Undecided=5	GPA	Years	Inches	Equally=3	fem = 2	Average = 2	Pounds	>2000=1
7	5	*	21	*	2	2	3	112	1
8	2	3.2	19	65	1	2	2	120	1
9	1	3.3	19	66	1	2	3	123	1
10	2	*	20	69	3	2	1	130	2
35	1	3.5	20	65	1	2	2	125	1
36	1	3	21	65	1	2	2	125	1
37	4	3.3	19	60	1	2	2	101	1
38	5	3	18	74	1	1	1	190	1
39	1	3	19	60	1	2	1	115	1
40	2	3	22	63	1	2	1	110	1
41	1	3.1	19	66	3	1	2	140	2
42	1	3.14	18	71	3	2	2	168	1
43									
44									
45									
46									
47									
48									
49									
50									
51									
52									

student data / order of questions–moon dia. / desir

When you want to return to a normal screen, just drag the split-screen cursor back to the top of the scroll bar, or select Remove Split from the Window menu. (By the way, you can also split the screen vertically. Just drag the vertical split indicator, which is located at the lower-right corner of the window, to the left. A quick way to split the screen in both directions at once: Select Split in the Window menu.)

Selecting Pieces of a Large Worksheet

Connected Range

Suppose you are working on a research project investigating the relationship between age and height. The range C4:D42 of the student data worksheet is relevant. Follow the instructions below to select just this range.

1. Click in the first cell of the range, in this case C4, which is in the top part of the split screen.
2. Hold down the SHIFT key and click on the cell D42 in the bottom part of the split screen.

You have now selected all of the cells from C4 to D42. (In the following illustration, some of the selected cells are hidden behind the split.)

	A	B	C	D	E
1	*Statistical Reasoning Class Survey*				
2	**Section 5, * means no response**				
3	College				More Likely?
4	Liberal Arts = 1	High		Height	HHTHTT=1
5	Business=2, Nursing = 3	School	Age in	in	HHHHHH=2
6	Science=4, Undecided=5	GPA	Years	Inches	Equally=3
7	5	*	21	*	2
8	2	3.2	19	65	1
9	1	3.3	19	66	1
35	1	3.5	20	65	1
36	1	3	21	65	1
37	4	3.3	19	60	1
38	5	3	18	74	1
39	1	3	19	60	1
40	2	3	22	63	1
41	1	3.1	19	66	3
42	1	3.14	18		3
43					

Disconnected Range

This time suppose you are interested just in heights and weights. In other words, the data of interest (including headings) is in the disconnected ranges D4:D42 and H5:H42.

1. First, click the mouse on cell D4. Then hold down the SHIFT key and click on cell D42, so the entire height column is selected.

	A	B	C	D	E
1	*Statistical Reasoning Class Survey*				
2	**Section 5, * means no response**				
3	College				More Likely?
4	Liberal Arts = 1	High		Height	HHTHTT=1
5	Business=2, Nursing = 3	School	Age in	in	HHHHHH=2
6	Science=4, Undecided=5	GPA	Years	Inches	Equally=3
7	5	*	21	*	2
8	2	3.2	19	65	1
9	1	3.3	19	66	1
35	1	3.5	20	65	1
36	1	3	21	65	1
37	4	3.3	19	60	1
38	5	3	18	74	1
39	1	3	19	60	1
40	2	3	22	63	1
41	1	3.1	19	66	3
42	1	3.14	18	71	3
43					

2. Now, hold down the CTRL key (the Apple key on a Mac) and click on H5, so that this cell as well as the column D range are both selected.

3. Finally, hold down the SHIFT key click on cell H42, and the two columns should both be selected as shown below.

	A	B	C	D	E	F	G	H	I
1	*Statistical Reasoning Class Survey*								
2	*Section 5, * means no response*								
3	College				More Likely?		Driving Skill		Moons Diameter
4	Liberal Arts = 1	High		Height in	HHTHTT=1	Gender	Better = 1		Estimate
5	Business=2, Nursing = 3	School	Age in	in	HHHHHH=2	male = 1	Worse = 3	Weight in	<2000=2
6	Science=4, Undecided=5	GPA	Years	Inches	Equally=3	fem = 2	Average = 2	Pounds	>2000=1
7	5	*	21	*	2	2	3	112	1
8	2	3.2	19	65	1	2	2	120	1
9	1	3.3	19	66	1	2	3	123	1
10	2	*	20	69	3	2	1	130	2
35	1	3.5	20	65	1	2	2	125	1
36	1	3	21	65	1	2	2	125	1
37	4	3.3	19	60	1	2	2	101	1
38	5	3	18	74	1	1	1	190	1
39	1	3	19	60	1	2	1	115	1
40	2	3	22	63	1	2	1	110	1
41	1	3.1	19	66	3	1	2	140	2
42	1	3.14	18	71	3	2	2	168	1

student data / order of questions - moon dia. / desir

2.3 HANDLING READ-ONLY FILES

The Excel workbooks saved on **STAT** are *read-only* files, which means that you cannot alter their contents in any permanent way. So, for example, if you make calculations in these workbooks, your work cannot be saved. This is a bit of a nuisance, but is nonetheless an important protection. Otherwise it would be too easy to inadvertently mangle or even delete needed data. Fortunately, there is an easy way around this problem. All you need to do is open the read-only workbook you are interested in and then open a new workbook so that you have two workbooks open at the same time, the read-only workbook and the new one. Then, select the data you want to work with and copy it from the read-only workbook to the new workbook. The new workbook will not be read-only, and so you will be able to edit it at will and save the changes whenever you like. When you save the new workbook for the first time, be sure to use the Save As command and give it a different name from that of the original read-only workbook.

2.4 WORKBOOKS AND WORKSHEETS

If you haven't already done so, open the **datach2.xls** workbook from **STAT** and then open a new workbook by clicking on the New Workbook button. Make a note of the number of the new workbook. You now have two open workbooks. The next section explains how to move from one to the other.

Moving Between Workbooks

1. Pull down the Window menu. At the bottom of the menu is a list of the open workbooks.

2. Select the workbook you wish to use, say **datach2.xls**, and it will appear on your computer screen.
3. To change back from **datach2.xls** to the new workbook, say **Book1**, pull down the Window menu again and select **Book1**. (Your workbook number may not be 1. Don't worry about the discrepancy. Just replace your number for 1 in the instructions that follow.)

Data Transfer

In this section, you will copy the gender and driving data from **datach2.xls** into a sheet of **Book1**.

1. Pull down the Window menu.
2. Select **datach2.xls** from the Window menu.
3. Select the Gender and Driving Skill data in columns F and G of **datach2.xls**: click in cell F3, hold down the SHIFT key and click in cell G42.
4. Pull down the Edit menu and choose Copy.
5. Select **Book1** from the Window menu.
6. Click in cell A1 of the new workbook and select Paste from the Edit menu.
7. Split the screen and, if you like, edit and format the headings so that you get a worksheet something like the following. Notice that the screen has been split, the headings formatted, and the data centered.

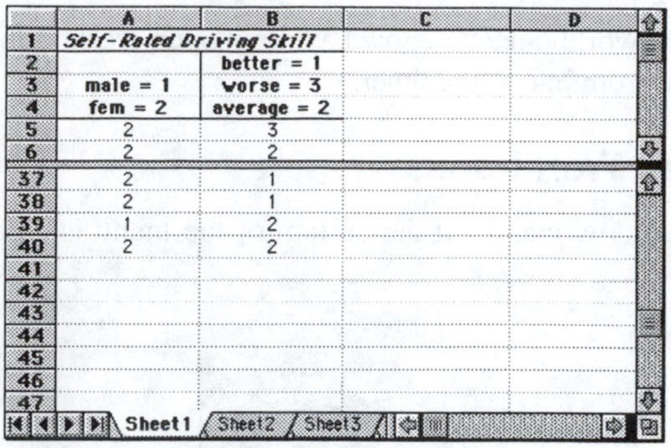

	A	B	C	D	
1	Self-Rated Driving Skill				
2		better = 1			
3	male = 1	worse = 3			
4	fem = 2	average = 2			
5	2	3			
6	2	2			
37	2	1			
38	2	1			
39	1	2			
40	2	2			
41					
42					
43					
44					
45					
46					
47					

Sheet1 / Sheet2 / Sheet3 /

Naming Worksheets in Excel 5 or 7

At the bottom of the window in boldface is the current name of the worksheet into which you pasted the data, in this case, **Sheet1**. Give the sheet a more descriptive name, say, "driving."

1.   Double-click on the **Sheet1** tab at the bottom of the window. The Rename Sheet dialog box will open.

2.   Type the name "driving" into the name slot.

Rename Sheet

Name
driving

OK

Cancel

3.   Click OK and the word **driving** will replace **Sheet1** in the worksheet tab.

Naming Worksheets in Excel 97

At the bottom of the window in boldface is the current name of the worksheet into which you pasted the data, in this case, **Sheet1**. Give the sheet a more descriptive name, say, "driving."

1.   Double-click on the **Sheet1** tab at the bottom of the window. The name will be highlighted as shown below.

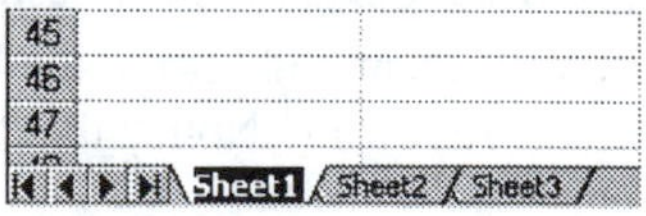

45	
46	
47	

Sheet1 / Sheet2 / Sheet3 /

2.   Type the word "driving" to give the sheet tab the new name.

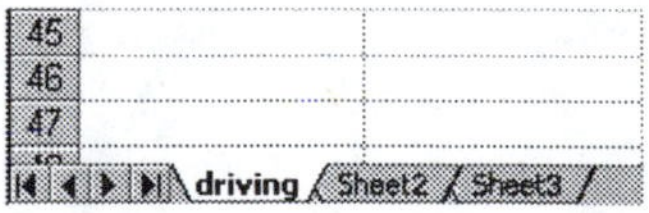

45	
46	
47	

driving / Sheet2 / Sheet3 /

Opening a New Worksheet

1. Click on the tab labeled **Sheet2**. A new worksheet will appear.
2. If no tab is visible, pull down the Insert menu and release on Worksheet.

Now, copy the GPA data from column B of **student data** into column A of this second worksheet and rename the worksheet "GPA." When you are done, your workbook will look like the illustration below.

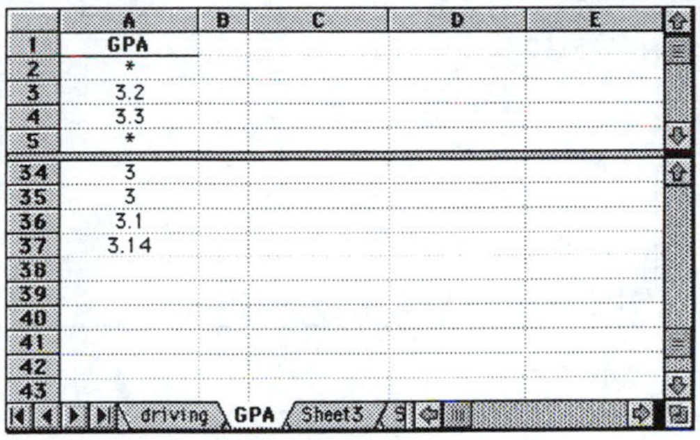

	A	B	C	D	E	
1	GPA					
2	*					
3	3.2					
4	3.3					
5	*					
34	3					
35	3					
36	3.1					
37	3.14					
38						
39						
40						
41						
42						
43						

driving \ **GPA** / Sheet3

Click on the tab labeled "driving" to return to that sheet.

Naming and Saving a Workbook

It might help to think of a worksheet as a chapter of a workbook rather than as a page. Worksheets have names just as the chapters of books do, and these worksheet names are different from the name of the workbook that holds them all. The "table of contents" for the workbook is the list of sheet names on the worksheet tabs while the "catalogue" of open workbooks is at the bottom of the Window menu.

Follow the instructions below to give the entire workbook the name "pracwb" for "practice workbook." (Any name is OK, so use another if you prefer.)

1. Pull down the File menu and release on Save As... .
2. Type the workbook name into the Save As dialog box, select a drive in which to save the file and click on the Save button.

Notice that the new name appears in the workbook title bar.

Deleting a Worksheet from a Workbook

At this juncture you probably don't need to remove any of the worksheets from **pracwb**, but for future reference here is how it is done. First of all, **do not close the workbook** by clicking, for example, on the upper-left corner of the window or by choosing Close from the File menu. You don't want to shut the whole workbook—you only want to remove one of its worksheets.

1. Click on the tab of the worksheet you wish to remove so that it becomes the active worksheet.
2. Pull down the Edit menu and click on Delete Sheet. The workbook now has one less sheet.

2.5 USING FUNCTIONS IN EXCEL

As you have probably learned in your study of statistics, one of the pitfalls of a survey as a method of collecting data is that people responding to survey questions sometimes hedge their replies. For example, when questioned about themselves, respondents tend to give the answers that put their own behavior and personal history in the best possible light. In this section, we will look at two survey questions, one that asked students to rate their own driving ability and another that asked students for their high school GPA. These are questions open to hedging. See if you can discern any evidence of hedging in the answers the students gave.

COUNT and COUNTIF

Make sure that the workbook that you just named **pracwb** is open and the **driving** worksheet visible. Column A lists the gender of the respondent where 1 stands for male and 2 for female. Column B contains each respondent's self-rating. An entry of 1 indicates a self-rating of "better than average" driving ability, an entry of 2 a self-rating of "average" driving ability, and 3 a self-rating of "worse than average" driving ability. Thus, the number 2 in cell A5 together with the 3 in cell B5 means that the first respondent listed was a woman who considered her driving ability to be "worse than average."

We begin with an analysis of all of the self-ratings ignoring, for the time being, the gender of the respondent. We will ask Excel to count the total number of respondents and then the number of times that each rating appears. Later, in the next section, an Excel function will be used to calculate the *average* or *mean* self-rating. Recall that the average or mean of a set of values is their total divided by their number.

1. Enter the headings "total number" in cell A42, "better" in cell A43, "average" in cell A44, "worse" in cell A45, and "mean rating" into cell A46. Add a lower border to A41:B41.

	A	B	C
1	*Self–Rated Driving Skill*		
2		better = 1	
3	male = 1	worse = 3	
4	fem = 2	average = 2	
5	2	3	
6	2	2	
38	2	1	
39	1	2	
40	2	2	
41			
42	total number		
43	better		
44	average		
45	worse		
46	mean rating		
47			

driving / GPA / Sheet3 / Sheet4 / Sheet

2. Now ask Excel to count the total number of ratings listed in column B: type the command =COUNT(B5:B40) into cell B42.

	A	B	C
1	*Self–Rated Driving Skill*		
2		better = 1	
3	male = 1	worse = 3	
4	fem = 2	average = 2	
5	2	3	
6	2	2	
38	2	1	
39	1	2	
40	2	2	
41			
42		=COUNT(B5:B40)	
43	better		
44	average		
45	worse		
46	mean rating		
47			

driving / GPA / Sheet3 / Sheet4 / Sheet

3. Press the ENTER key and the total number of respondents, 36, is returned in cell B42.

	A	B	C
1	*Self-Rated Driving Skill*		
2		better = 1	
3	male = 1	worse = 3	
4	fem = 2	average = 2	
5	2	3	
6	2	2	
38	2	1	
39	1	2	
40	2	2	
41			
42	total number	36	
43	better		
44	average		
45	worse		
46	mean rating		
47			

`|◀|◀|▶|▶|\ driving / GPA / Sheet3 / Sheet4 / Shee▏`

4. Of those 36 respondents, how many rated themselves as "better than average" drivers? In other words, how many 1's are there in the range B5:B40? Type the command =COUNTIF(B5:B40,1) into cell B43 as shown next.

	A	B	C
1	*Self-Rated Driving Skill*		
2		better = 1	
3	male = 1	worse = 3	
4	fem = 2	average = 2	
5	2	3	
6	2	2	
37	2	1	
38	2	1	
39	1	2	
40	2	2	
41			
42	total number	36	
43	better	=COUNTIF(B5:B40,1)	
44	average		
45	worse		
46	mean rating		
47			

`|◀|◀|▶|▶|\ driving / GPA / Sheet3 / Sheet4 ▏|◁|▦▏`

Press the ENTER key and Excel provides the count given in the illustration below: 17 of the 36 students thought their driving ability "better than average."

	A	B	C
1	*Self-Rated Driving Skill*		
2		better = 1	
3	male = 1	worse = 3	
4	fem = 2	average = 2	
5	2	3	
6	2	2	
37	2	1	
38	2	1	
39	1	2	
40	2	2	
41			
42	total number	36	
43	better	17	
44	average		
45	worse		
46	mean rating		
47			

driving / GPA / Sheet3 / Sheet4

5. Now, count the number of students who gave each of the other two responses. Enter the command =COUNTIF(B5:B40,2) into cell B44 and =COUNTIF(B5:B40,3) into cell B45. The result is shown below.

	A	B	C
1	*Self-Rated Driving Skill*		
2		better = 1	
3	male = 1	worse = 3	
4	fem = 2	average = 2	
5	2	3	
6	2	2	
37	2	1	
38	2	1	
39	1	2	
40	2	2	
41			
42	total number	36	
43	better	17	
44	average	16	
45	worse	3	
46	mean rating		
47			

driving / GPA / Sheet3 / Sheet4

So, 16 students rated themselves about average in driving ability while only 3 were willing to admit to "worse than average" driving skills.

About Functions

Since this is our first encounter with Excel functions, other than the use of the Sum button, let's take a second look at the various features of the formula =COUNTIF(B5:B40,1). The equal sign is important; it tells Excel that it must perform a calculation. The expression COUNTIF is a special Excel command meaning "look through a range of values and count those that are equal to a specified value." This command is followed by parentheses that contain two entries, the range where Excel is to make the count, followed by the specified value for which it must look. Any command, such as COUNTIF, that requires "input" is called a *function*, and the pieces of input information (in this case A5:A40 and 1) are called the *arguments* of the function. The COUNT function, you recall, has just a single argument, the range of cells where all entries are to be counted.

The AVERAGE Function

Select cell B46 and enter the command =AVERAGE(B5:B40), which will produce the mean or average rating that students gave themselves. It is 1.61 as shown below.

	A	B	C
1	*Self-Rated Driving Skill*		
2		better = 1	
3	male = 1	worse = 3	
4	fem = 2	average = 2	
5	2	3	
6	2	2	
37	2	1	
38	2	1	
39	1	2	
40	2	2	
41			
42	total number	36	
43	better	17	
44	average	16	
45	worse	3	
46	mean rating	1.61	
47			

driving / GPA / Sheet3 / Sheet4

Notice that students rated themselves very highly just as survey theory suggests. Virtually no one admitted to "worse than average" driving ability. Now, it may be that these students are all excellent drivers, but the fact of the matter is that when survey questions ask people to rate themselves on driving ability or anything else, hardly anyone is ever "worse than average," at least by their own estimation!

Sorting Data Using the IF Function

Is there a difference in the way that men and women rated themselves? Was one gender harder on itself (perhaps more honest?) than the other, or were the responses about the same for both? To answer these questions, the data must be sorted into female and male responses. Let's say we decide to extract just the female ratings and place them in column C. Which responses are these? As an example, look at the entry in cell B38. It is a 1 indicating a self-rating of "better than average." Was this a male response or a female response? Notice that the entry in cell A38 is a 2 which tells you that the B38 rating was given by a female. On the other hand, line 39 contains a male response as the entry in A39 is 1. So, what is needed is a command that will cause Excel to first look at the column-A entry and, if it is a 2, place the corresponding column-B entry in the column C. Excel's IF function is perfect for this task.

1. Begin by typing the command =IF(A5=2,B5," ") into cell C5 as shown next.

	A	B	C	D
1	*Self-Rated Driving Skill*			
2		better = 1		
3	male = 1	worse = 3		
4	fem = 2	average = 2		
5	2	3	=IF(A5=2,B5," ")	
6	2	2		
37	2	1		
38	2	1		
39	1	2		
40	2	2		
41				
42	total number	36		
43	better	17		
44	average	16		
45	worse	3		
46	mean rating	1.61		
47				
	driving / GPA / Sheet3 / Sheet4 /			

This command gives the following instruction to Excel: if the contents of cell A5 is a 2, display the contents of cell B5, otherwise display nothing. (Two quotation marks separated by a space means "display nothing.")

2. Press the ENTER key. The number 3 appears in cell C5 just as it should.

3. Copy the IF command down column C: select cell C5 and move the cursor to the lower right corner of the cell. When it becomes a black cross, double-click. The result is shown next.

	A	B	C	D
1	*Self-Rated Driving Skill*			
2		better = 1		
3	male = 1	worse = 3		
4	fem = 2	average = 2		
5	2	3	3	
6	2	2	2	
37	2	1	1	
38	2	1	1	
39	1	2		
40	2	2	2	
41				
42	total number	36		
43	better	17		
44	average	16		
45	worse	3		
46	mean rating	1.61		
47				
	driving / GPA / Sheet3 / Sheet4 / Sheet5 /			

Check to make sure that Excel did the correct thing. Scroll down column C. The ratings displayed should be only those corresponding to female respondents. All cells like C39, for example, contained in a male row should be blank.

4. Now that the female ratings have been sorted out, we can count the total number of them, the number of female responses in each rating category, and the average female rating. This is quickly accomplished by copying the formulas in B42:B46 to the range C42:C46. Click in cell B42, then hold down the SHIFT key and click in cell B46.

Move the cursor to the lower right corner of cell B46. When it becomes a black cross, press and hold the mouse button and drag across to cell C46. The result is displayed next. Notice that the headings "All" and "Female" were added to make the column categories clear.

	A	B	C	D
1	*Self-Rated Driving Skill*			
2		better = 1		
3	male = 1	worse = 3		
4	fem = 2	average = 2		
5	2	3	3	
6	2	2	2	
37	2	1	1	
38	2	1	1	
39	1	2		
40	2	2	2	
41		All	Female	
42	total number	36	25	
43	better	17	11	
44	average	16	11	
45	worse	3	3	
46	mean rating	1.61	1.68	
47				

driving / GPA / Sheet3 / Sheet4 / Sheet5 / S↔

Now, sort out the male ratings and make the same set of calculations for them.

1. Enter the formula =IF(A5=1,B5," ") into cell D5. The output should be a blank cell.

2. Move the cursor to the lower right corner of cell D5 and when it becomes a black cross, double-click to copy the IF command down column D.

3. Copy the formulas from C42:C46 to D42:D46 using a black cross drag. Your worksheet should then look like the one shown below. Remember to type in the column label in cell D41.

	A	B	C	D
1	*Self-Rated Driving Skill*			
2		better = 1		
3	male = 1	worse = 3		
4	fem = 2	average = 2		
5	2	3	3	
6	2	2	2	
37	2	1	1	
38	2	1	1	
39	1	2		2
40	2	2	2	
41		All	Female	Male
42	total number	36	25	11
43	better	17	11	6
44	average	16	11	5
45	worse	3	3	0
46	mean rating	1.61	1.68	1.45
47				

driving / GPA / Sheet3 / Sheet4 / Sheet5 /

Notice that the average female rating is lower than the average male rating and that not a single man rated his driving ability as "worse than average." Maybe women are harder on themselves—or more honest—than men, or perhaps men are better drivers! You will have the chance to investigate this issue further using a much larger data set in the exercises.

2.6 FUNCTION WIZARD AND PASTE FUNCTION

Excel has many, many built-in commands like COUNT, COUNTIF, IF, and AVERAGE that can be used to automate spreadsheet calculations. If you don't know the exact spelling for a command or you can't remember how the arguments work, you can use a special Excel utility called either Function Wizard or Paste Function, depending on the version of Excel you are using, which can help you find and implement the function you need for a particular task.

The use of these utilities is always optional. When you know the command you want to use and how it works, there is no need for them. For those times when they might be helpful, follow the steps of the procedure below in which the median high school GPA is calculated for the data in **GPA**. (The *median* of a data set, you recall, is the middle value. So, for example, the median for the set of numbers 10, 20, 30, 40, 50 is 30.)

Function Wizard in Excel 5 and 7

Click on the **GPA** tab at the bottom of **pracwb**. We will use the Paste Function dialog box to place the median GPA in cell D1 of the worksheet.

1. Select cell C1. Type "median GPA" and select cell D1.
2. Click on the Function Wizard button, shown below.

3. The Function Wizard dialog box will appear.

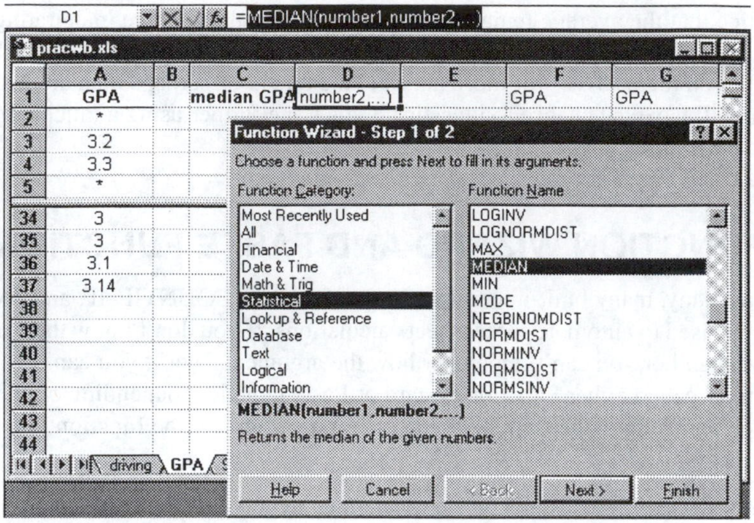

4. Make sure that "Statistical" is selected in the Function Category window and that MEDIAN is selected in the Function Name window. Press the Next button. The dialog box for the second step of Function Wizard appears.

5. Type the range of numbers for which you wish the median in the **number 1** slot. (Alternatively, you can click and drag the dialog box out of the way, so that the desired range is visible; then highlight the range with the mouse.)

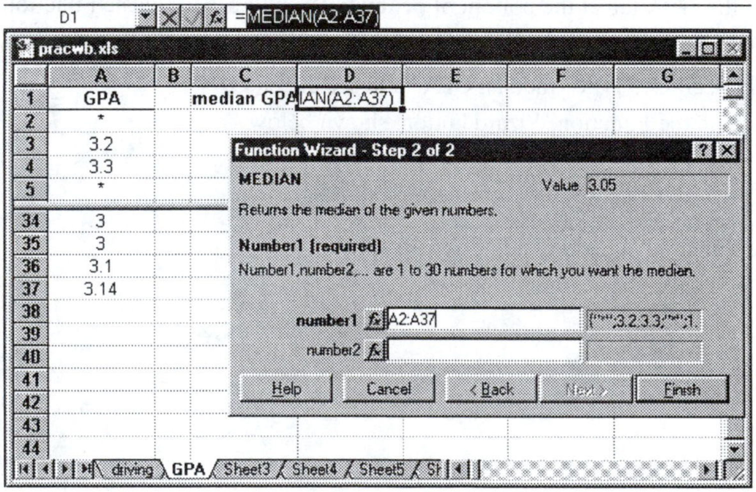

Before going on, take a minute to look at the rest of the dialog box and at your worksheet. Notice that Excel has listed the GPA values to the right of the **number1** range. This permits you to double-check that the range you entered is really the one you want. The top right of the dialog box contains the median of the values displayed, in this case 3.05. Now look at the worksheet. Both cell D1 and the title bar contain the MEDIAN function with the range A2:A37 as its argument.

6. Press the Finish button to enter the numerical median, 3.05, in cell D1.

In the exercises the self-reported GPAs of all the respondents will be studied. You will be able to decide for yourself whether you think there is evidence of GPA inflation.

Paste Function in Excel 97

Click on the **GPA** tab at the bottom of **pracwb**. We will use the Paste Function dialog box to place the median GPA in cell D1 of the worksheet.

1. Select cell C1. Type "median GPA" and click in cell D1.
2. Click on the Paste Function button shown below.

3. The Paste Function dialog box will appear.

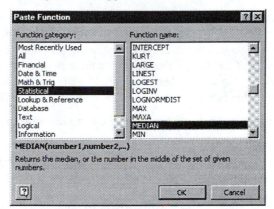

4. Make sure that "Statistical" is selected in the Function Category window and that MEDIAN is selected in the Function Name window. Press the Next button. The dialog box for the second step of Function Wizard appears.
5. Type the range of numbers for which you wish the median in the **number 1** slot. (Alternatively, you can click and drag the dialog box out of the way, so that the desired range is visible; then highlight the range with the mouse.)

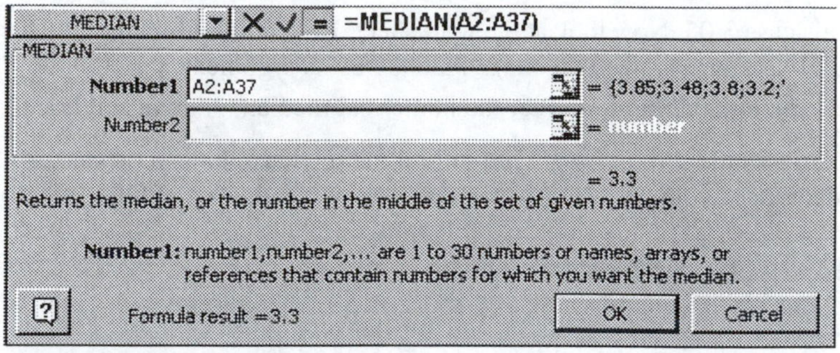

6.   Press the Finish button to enter the numerical median, 3.05, in cell D1.

Later in the exercises the GPAs of many more respondents will be investigated for evidence of GPA inflation.

2.7 THE FIVE-NUMBER SUMMARY

The average and the median are called *measures of central tendency* because each provides a middle figure for the data. It is also important to have an idea of how the data spreads to the right and left of its center values. One device often used to depict spread is the *five-number summary* of the data. A five-number summary consists of the minimum data value, the first quartile (the number below which 25% of the data lies), the median, the third quartile (the number below which 75% of the data lies), and the maximum data value.

The Excel commands listed below will calculate a five-number summary of a data set.

Excel Commands for the Five-Number Summary	
minimum	=MIN(*data range*)
first quartile	=QUARTILE(*data range*,1)
median	=MEDIAN(*data range*)
third quartile	=QUARTILE(*data range*,3)
maximum	=MAX(*data range*)

Open the worksheet **High School GPA** on **datach2.xls**. The picture below shows this worksheet with the five-number summary for the data entered in the range C242:C246. The first quartile command is displayed in the formula bar and the first quartile value, 3, in cell C245.

		=	=QUARTILE(A2:A248,1)	
	C245	▾		
	A	**B**	**C**	**D**
1	**GPA**			
2	3.85			
3	3.48			
241	3.2	Five Number Summary		
242	3.5	maximum	34	
243	4	third quartile	3.5	
244	3.56	median	3.2	
245	3.3	first quartile	3	
246	*	minimum	0	
247	4.2			
248	3.83			
249				

Make these same calculations in your own worksheet.

The five-number summary clearly reveals *outliers* in the data. They probably result from data entry errors. After all, a high school GPA of 34 is impossible, and one of 0 highly unlikely for a university student. Knowledge of outliers like these is extremely important in analyzing data because their presence can sharply skew many statistical measures such as the average. In the exercises you will work with a variation of the five-number summary that eliminates outliers of this type and so helps to give a better idea of the real spread of the data.

EXERCISES

In responding to the questions below, make all of your calculations in a worksheet and place your written answers in text boxes. Print only the required calculations and the boxes. This means that you must select just that portion of your worksheet containing these items—it may be necessary to reposition the text box—*before* issuing the print command, and then you must be careful to choose "Print Selection" from the Print dialog box. (Don't forget to type your name in the print region.) An irresponsible Print command means that dozens of unnecessary pages of raw data will be spewed out by your printer. This wastes machine time and trees.

Exercise 2.1 Open **datach2.xls** and create the workbook **pracwb** as described in Section 2.4.
  (a) Reproduce the sorting and analysis of the driving self-rating done in the chapter. Select the resulting table and print it.
  (b) Construct a table containing the mean or average, the median, and the mode of the GPA data and print it. (The Excel command =MODE(*data range*) calculates the mode.)

Exercise 2.2 Open the worksheet entitled **driving-gender**. Read the text box that describes it.

(a) Follow the instructions in the chapter to produce a table of counts, ratings, and average rating for all respondents.

(b) Discuss any evidence you see that respondents inflated their actual driving ability.

Exercise 2.3 Open the worksheet entitled **driving-gender**. Read the text box that describes it.

(a) Calculate the average, median, and mode for all students. The Excel command for mode is =MODE(*data range*).

(b) Which measure of central tendency—mean, median, or mode—do you think best represents this data and why? Type your answer in a text box.

(c) Open a text box and write a short paragraph discussing the problems inherent in self-rating as a data-gathering tool. Compose a survey question or two that might provide a more valid measure of driving ability.

Exercise 2.4 Open the worksheet entitled **driving-gender**. Read the text box that describes it.

(a) Sort the data into male and female ratings and produce a table of counts, ratings, and average rating for each.

(b) Briefly discuss any differences you see in the pattern of ratings.

Exercise 2.5 Open the worksheet **High School GPA**. Read the text box that contains a description of this data. Notice that the worksheet also contains a historical list of the actual average high school GPAs of entering students at the institution where the survey was made.

(a) Calculate the average, median, and mode of the GPA data.

(b) Why might you expect the self-reporting of GPA to be more accurate than the self-rating of driving ability?

(c) Compare the average, median, and mode of the GPA data with the historical record. (Keep in mind that most, but not all, of the students responding to the survey were freshmen or sophomores.) Is there any indication of "hedging" in the self-reported data or does it seem close to what might be expected given the historical record?

(d) If you do see a discrepancy between the self-reported GPAs and the historical data, can you think of an explanation for it other than false self-reporting?

Exercise 2.6 Open the worksheet **order of question-moon diameter**. Read the text box that contains a description of this survey data.

(a) Compute the mean, median, and mode of the moon diameter guesses for all those students for whom the anchor question read "Is the diameter of the moon more or less than 1000 miles?"

(b) Compute the mean, median, and modal moon diameter guesses for all those students for whom the anchor question read "Is the diameter of the moon more or less than 2000 miles?"

(c) Did the order of the questions—asking students first if the diameter of the moon was more or less than a specified value and then asking them to guess the diameter of the moon—have a biasing effect on the moon diameter guesses they gave? Explain in a text box.

(d) Explain why the averages calculated in (a) and (b) above are so much larger than the median and modal values. Which measure of central tendency is the best one to use for this data and why? Again, answer in a text box.

Exercise 2.7 Open the worksheet called **bias-honesty questions**. Read the text box that contains a description of the data in the worksheet.
(a) What proportion of the students who responded to the question
    If you found a wallet with $20 in it, would you keep the money?
    said they would *not* keep the money?
(b) What proportion of the students who responded to the question
    If you found a wallet with $20 in it, would you do the honest thing and return
        the money?
    said they would return the money?
(c) Did the prompt "do the honest thing" seem to have an effect on the responses? If so, did it make a lot of difference?

Exercise 2.8 Open the worksheet **High School GPA**.
(a) Calculate a five-number summary for the data in the worksheet.
(b) Now calculate an adjusted five-number summary that eliminates the outliers: replace the maximum data value with the 90th percentile (the number below which 90% of the data lies) and the minimum by the 10th percentile (the number below which 90% of the data lies). The 90th percentile is calculated with the command
    $$=PERCENTILE(data\ range, 0.90)$$
and the 10th percentile with
    $$=PERCENTILE(data\ range, 0.10)$$

Exercise 2.9 Open the **exercise-gender** worksheet.[1] In thinking about this data keep in mind that a low pulse rate is considered healthy since it indicates that the heart is beating more slowly and so not working as hard.
(a) Use the IF command to sort the pulse rate data into two columns, one for exercisers and one for nonexercisers.
(b) Calculate the average pulse rate for both groups.
(c) Calculate the five-number summaries for the pulse rate of exercisers and non-exercisers.
(d) Open a text box and write a discussion of any differences in pulse rates for exercisers and nonexercisers. Which group has the generally healthier rate?
(e) Use the IF command to sort the pulse rate data into two columns, one for men and one for women, and find the average pulse rate of each. Calculate five-number summaries for each.
(f) Open a text box and explain why it would be a good idea to further sort the exercisers and nonexercisers by gender before making any claims about the beneficial effects of exercise on pulse rate.

---

[1]Data provided by Professor Jessica M. Utts, University of California, Davis

WHAT TO HAND IN: For each assigned exercise, print just your calculations and text boxes. DO NOT PRINT THE ENTIRE WORKBOOK.

Histograms and Boxplots

3

In This Chapter...

- Drawing boxplots
- Moving and resizing plots
- Plotting histograms

The best way to quickly get a feel for the distribution of data values is to draw a picture of it in the form of a histogram or a boxplot. Unfortunately, Excel does not have a built-in boxplot program. It can draw a histogram, but the procedure is difficult to use—not at all user-friendly. We have addressed both problems by writing programs that automate these drawing tasks. A program of this type, which is designed to do a special job, is called a *macro*.

Since the histogram and the boxplot macros are not a built-in part of Excel, the following instructions will not work on your computer until you have first installed these macros in your Excel Library file and then selected them from your list of add-in tools. See Appendix B for instructions. (Note: if you are a student and using a university or college computer lab, these macros were probably installed by your instructor.)

3.1 BOXPLOTS

Reading a Boxplot

The following boxplot shows the average annual salaries, measured in thousands of dollars, earned by professors at 50 institutions of higher learning in the United States.

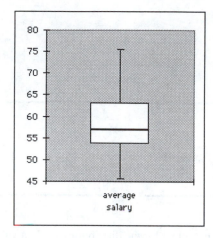

The box portion of a boxplot runs from the 25th to the 75th percentile. In other words it covers the interquartile range of the data. Reading from the bottom edge of the box directly across to the vertical scale, you can see that the low salary in this range is about $54,000. The high salary is $63,000. So, the middle 50% of institutions covered by the data pay their professors, on average, between $54,000 and $63,000 per year.

The heavy line drawn across the interquartile box indicates the median salary, here about $57,000.

The vertical lines, sometimes called *whiskers*, are drawn from the top and bottom of the box to the largest and smallest data salary values. In this case, $46,000 and $76,000 are the minimum and maximum average salaries, respectively, reported by the 50 institutions. When data sets are large or are known to contain outliers, extreme data values can be eliminated before the plot is drawn so that the whiskers extend to, say, the 5th and 95th percentiles instead of to the endpoints of the range.

The boxplot makes it clear that the salary data is what is called *skewed to the right*: the interval of salary values lying above the median ($57,000 to $76,000) is longer (almost twice as long) as the interval ($46,000 to $57,000) lying below the median.

Drawing a Boxplot

Begin by opening the workbook entitled **datach3.xls** from **STAT**. Select the worksheet entitled **prof salary**.[1] Split the screen so that lines 1 and 51 are both visible. This section contains instructions for producing the average salary boxplot shown above.

1. Select the data to be plotted. Click on the heading "average salary" in cell B1, hold down the SHIFT key and click in cell B51. Your worksheet should look like the picture below.

	A	B	C	D	E
1	University	average salary	asst prof salary	assoc prof salary	full prof salary
2	DUKE	64.47	46.1	57.5	83
3	VANDERBILT	59.2	42.5	49.7	78.9
4	WASHINGTON U	58.77	43.8	51.4	75.4
5	TULANE	55.94	41.5	50.8	70.2
6	CAL TECH	75.54	56.4	70	93.3
42	U.S.C.	61.67	46.1	55.2	77.9
43	KANSAS	45.76	35.9	41.8	55.9
44	U.C.L.A.	59.16	42.9	51.5	76.9
45	U. OF WASH.	53.15	41.3	46.9	66.7
46	OREGON	45.75	35.3	42	56.1
47	CAL SAN DIEGO	59.3	43	50.5	75
48	CLARK	52.06	40	46.3	65.3
49	CATHOLIC U	48.85	37.2	44	61
50	N.Y.U.	66.31	51	56.8	85.1
51	COLUMBIA	63.22	43	55.8	83.4
52					

Exercise Gender \ **prof salary** \ test data \ age g

2. Pull down the Tools menu and select Boxplot.[2] The Boxplot: Select Data dialog box will appear.

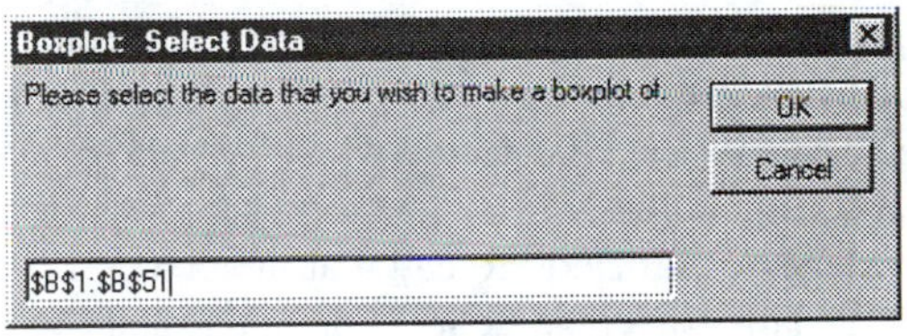

```
Boxplot: Select Data                          [x]
Please select the data that you wish to make a boxplot of.    OK
                                                           Cancel

$B$1:$B$51
```

This dialog box indicates the range of cells Excel will use to produce the boxplot. The range shown $B$1:$B$51 is correct. If there were an error, you could change it now before going on.

Since no corrections have to be made, click the OK button.

---

[1] Retrieved from the Data and Story Library (DASL) website. Reference: "Faculty Compensation and Benefits." (1993, April). Ohio State University.
[2] Boxplot can also be launched with a keyboard command—hold down the SHIFT key and the CTRL key and press the B key.

3. The Boxplot Parameters dialog box is displayed next.

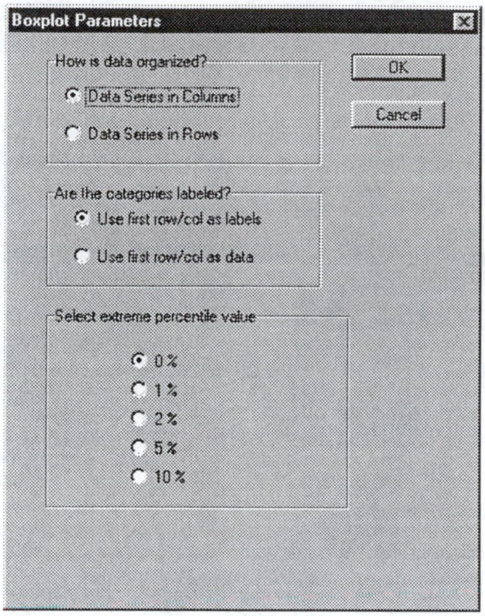

It asks for three entries.

(a) **How is data organized?** In this case, the salary figures were in column B of the worksheet. The answer that indicates this "Data Series in Columns" is selected, so you can go on to the next question. Had the data been in a row of the worksheet, you would have checked the row response at this point. The third response, "Treat all Data as Data Series," is selected when the data to be plotted is contained in a block of rows and columns.

(b) **Are the categories labeled?** Recall that the first cell (B1) of the data selected contained the label "average salary" and not a salary value. So, the first response "Use first row\col as labels" is correct.

(c) **Select extreme percentile value.** Your response to this instruction will determine the endpoints of the boxplot whiskers. When 0% is checked, as it is here, the whiskers will extend down to the 0th percentile and up to the 100th percentile. If you were to check 5% instead, the whiskers would go to the 5th and 95th percentiles. Leave this setting as it is.

4. When you have answered all of the questions, click the OK button. Excel will embed the plot in a numbered boxplot sheet.

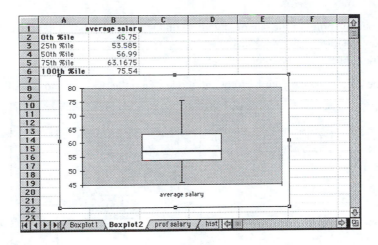

In the illustration, the name of the sheet is **Boxplot2**. This sheet can be renamed or deleted just like any other sheet in the workbook. Notice that the upper left corner of the boxplot sheet contains relevant percentile values for the data. The 0th and 100th percentile are shown in bold face to indicate the extreme values to which the whiskers are drawn. Had you set the extreme value parameter to 5%, these values would be the 5th and the 95th percentiles.

Some Boxplot Editing Techniques

It is easy to alter the size of a boxplot or to change its position in a worksheet.

How to Select a Boxplot

A boxplot must be *selected* before it can be edited. A selected boxplot has a thin border containing eight rectangular handles as in the picture above. Excel automatically selects a boxplot when it embeds it in a boxplot sheet. If your boxplot does not show these handles, click once inside the plot, and they will appear. If you mistakenly click on the boxplot more than once, you can recover the situation. Just click in the worksheet outside the boxplot until the border of the boxplot becomes a thin line and then click in the boxplot once to obtain the handles.

How to Change the Size of a Boxplot

1. If the boxplot is not already selected, click on it once to select it.
2. Place the point of the arrow on one of the black handles. Click and drag the mouse. The size and shape of the boxplot will change depending on the direction of the drag.
3. Click outside the boxplot to deselect it.

How to Reposition a Boxplot

1. Be sure that the boxplot is selected.

2. Place the point of the cursor anywhere on the boxplot. Click and drag the boxplot to a new position in the worksheet.

3. Click outside the boxplot to deselect it.

Practice a little: resize your boxplot and move it into the position shown below.

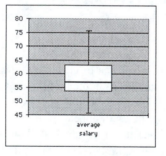

Adding Gridlines

You can draw horizontal gridlines across the background of the boxplot.

1. Select the boxplot.

2. Hold down the SHIFT key and the CTRL key and press the G key.

The keystroke sequence SHIFT-CTRL-G can also be used to remove gridlines from a selected plot. A command or button that has this effect, that turns a feature on or off, is called a *toggle*.

How to Transfer a Boxplot

It is a simple matter to move a boxplot from the boxplot sheet to any other sheet in the workbook.

1. Select the boxplot by clicking on it once.

2. Pull down the Edit menu and release on Copy.

3. Open the worksheet to which you wish to move the boxplot.

4. Click in the new worksheet.

5. Pull down the Edit menu and release on Paste.

How to Delete a Boxplot from a Worksheet

1. Select the boxplot by clicking on it once.

2. Press the DELETE key.

A Quick Way to Remove an Entire Worksheet from a Workbook

1. Click on the tab for the worksheet you wish to remove.

2. Hold down the SHIFT key and the CTRL key and press the K key ("K" for *kill*).

The editing techniques presented here are really just the minimum needed to produce readable boxplots. Many other options are available in Excel such as changing the axis scale or adding titles to the chart. For a fuller treatment of chart editing refer to Section 6.1.

3.2 HISTOGRAMS

Another method of picturing data sets is by means of a specialized bar chart called a *histogram*. A histogram of the salary data is displayed below.

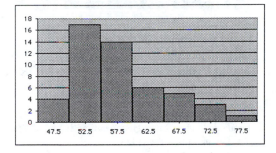

Reading a Histogram

In the histogram above, the first column on the left is labeled 47.5. Its height is 4. This does *not* mean that the salary data set contained 4 entries equal to $47,500; it means that there are 4 salary figures *between* $45,000 and $50,000 in the salary data. These dollar figures correspond to the endpoints of the column's base, 45 and 50. These numbers are not shown on the horizontal axis, but they can be deduced from the midpoint values. The next column's height is 17, meaning that Excel found 17 salary values in the interval $50,000 \leq salaryvalue < $55,000. This column is labeled 52.5, again indicating the midpoint of the corresponding salary interval. The remaining columns are read in a similar manner. The intervals over which the columns are drawn are called *bins*, so in the example the first bin runs from 45 to 50, the second from 50 to 55, the third from 55 to 60, and so forth. In each case the *bin length* is 5. In our example, these bins *include their left boundary*. This means, for example, that a data value of $50,000 would be counted in the second bin, but one of $55,000 would be counted in the third bin, which covers the interval $55,000 \leq salaryvalue < $60,000$. The importance of these technicalities will become clearer to you as you gain more experience with histograms.

Drawing a Histogram

In this section you will produce the histogram of the salary data discussed above. If you haven't done so already, begin by opening the workbook entitled **datach3.xls** from **STAT**. Select the worksheet entitled **prof salary**. Split the screen so that rows 1 and 51 are both visible.

1. Select the data to be plotted. Do not include the heading. Click on cell B2, hold down the SHIFT key and click in cell B51.

2. Pull down the Tools menu and select Smart Histogram.[3] The Select Data dialog box will appear.

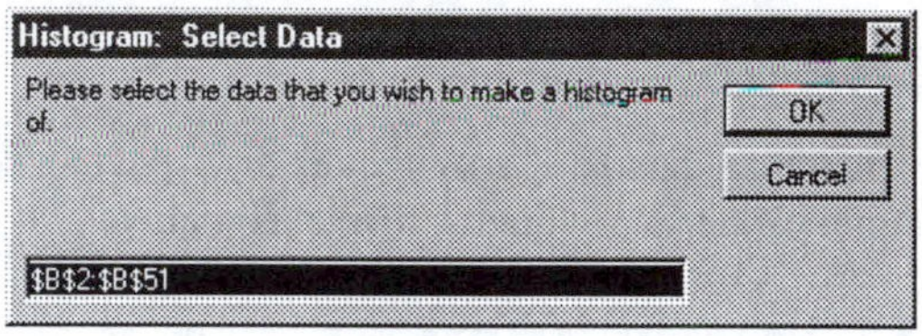

This dialog box indicates the range of cells Excel will use to produce the boxplot. The range shown $B$2:$B$51 is correct. If there were an error, you could change it now before going on. Since no corrections are needed, click the OK button.

3. The Histogram Parameters dialog box appears next.

---

[3]In some versions of Excel you can launch Smart Histogram with a keyboard command—hold down the SHIFT key and the CTRL key and press the H key.

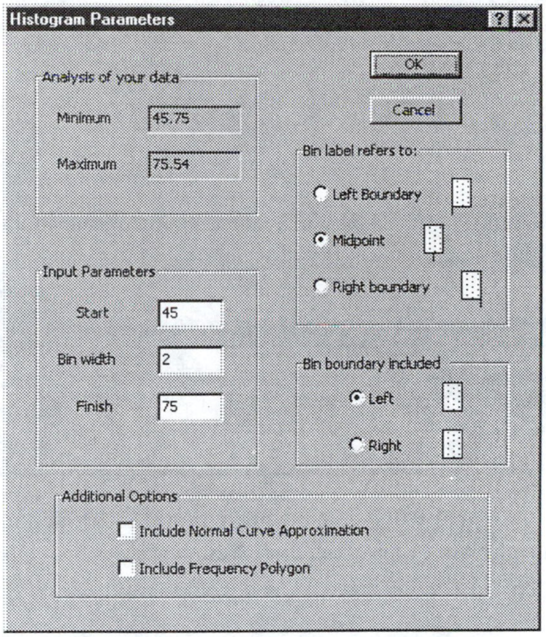

Under **Analysis of your data** the dialog box displays the minimum and maximum values found in the data. Notice that these values are paled out, meaning that you cannot change them. They are shown simply for your information.

4. Leave **Bin label refers to** set at **Midpoint**. This setting controls the numbers which are printed on the horizontal axis of the histogram. In the salary average histogram these were the midpoints of each bin.

5. It sometimes happens that a data value falls exactly at the point where one bin ends and the next begins. Since we do not want data counted in two bins, we need to decide where such values should go. Setting **Bin boundary included** to **Left** means that a value will be counted in a bin in the case that it falls exactly on the left boundary of the bin but will be placed in the next bin when it falls exactly on the right boundary. Since the salary data histogram was constructed with the left bin boundary included, leave this option checked.

The **Input Parameters** need to be altered. The current values represent Excel's best guess at appropriate **Start**, **Finish**, and **Bin width** values for the histogram. The Start and Finish values refer to left boundaries of the first and the last bin. These values are correctly set at 45 and 75, but the bin width needs to be changed to 5. Click in the **Bin width** box and type in the number **5**, as shown next.

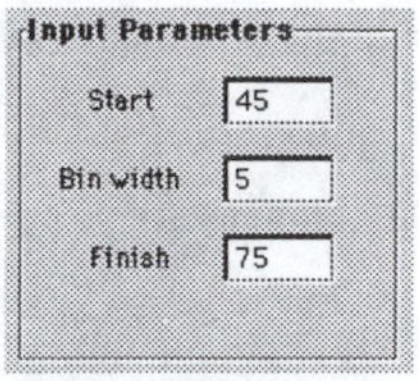

```
┌─Input Parameters──────┐
│                        │
│    Start    [45   ]    │
│                        │
│   Bin width [5    ]    │
│                        │
│   Finish    [75   ]    │
│                        │
└────────────────────────┘
```

6. Click the OK button and Excel will embed the histogram in a numbered histogram sheet and at the same time ask if you wish to keep the sheet. The query box is shown next.

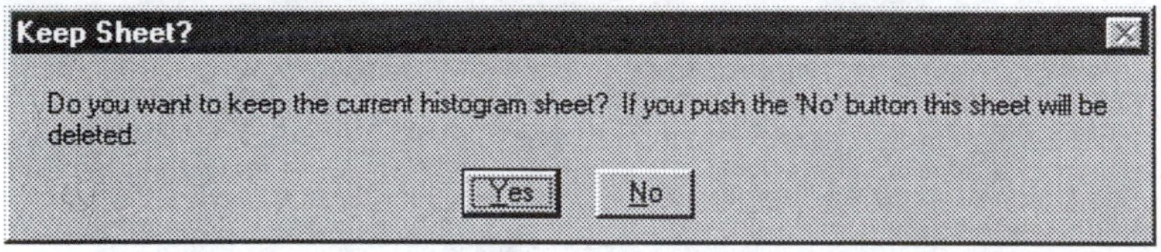

```
┌─────────────────────────────────────────────────────────────────────┐
│ Keep Sheet?                                                      [X]  │
├─────────────────────────────────────────────────────────────────────┤
│  Do you want to keep the current histogram sheet? If you push the 'No' button this sheet will be │
│  deleted.                                                             │
│                                                                       │
│                        [ Yes ]      [ No ]                            │
└─────────────────────────────────────────────────────────────────────┘
```

If you click No, Excel will delete the histogram sheet, which means you will lose the histogram just drawn. In this case, click Yes.

7. A second query box opens that asks you if you wish to draw a new histogram of the same data.

```
┌────────────────────────────────────────────────────┐
│ Another Histogram?                            [X]   │
├────────────────────────────────────────────────────┤
│  Do you want to draw another histogram of the same data? │
│                 [ Yes ]     [ No ]                  │
└────────────────────────────────────────────────────┘
```

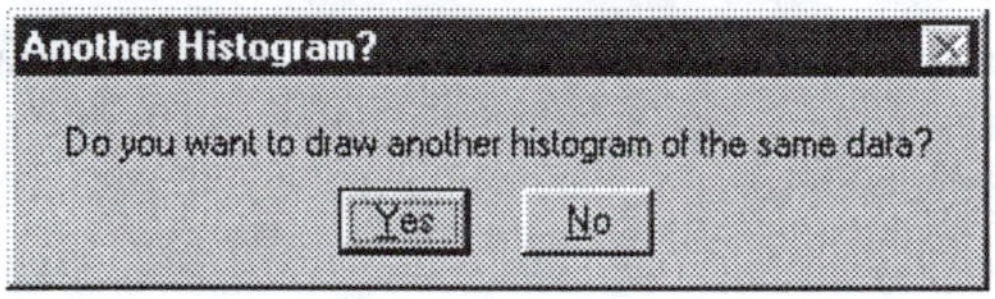

If you click Yes, Excel will return you to the Histogram Parameter box where you will be able to select new entries for the bin boundary inclusion, start and finish values, and bin width. In this case, click No. Excel will close the query box and return to the histogram sheet as pictured next.

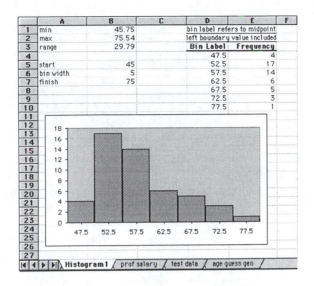

Editing a Histogram

You can reposition the histogram in its histogram sheet, change its size or shape, move it to a new sheet, or add gridlines to the background. The instructions for doing so can be found in the earlier section entitled "Some Boxplot Editing Techniques." As you read that section, just mentally replace the word "boxplot" with the word "histogram."

EXERCISES

Data for these exercises can be found in **datach3.xls** in **STAT**.

Exercise 3.1 Produce the professor salary data boxplot discussed in Section 3.1. Print it together with the percentile data appearing in the upper left corner of the worksheet.

Exercise 3.2 Produce the professor salary data histogram discussed in Section 3.2. Select the portion of the histogram sheet containing the chart, the frequency table, and the parameter values, and print this selection.

Exercise 3.3 Open the worksheet entitled **test data**. It is possible to draw many different histograms for the same set of data. Some will do a better job than others in describing the distribution. In this exercise you will have the chance to compare several histograms drawn for the same hypothetical data set.

(a) Examine the test data and, without drawing a boxplot or histogram, try to describe the distribution of the data. Is it skewed or symmetric? are there outliers?
(b) Draw the default histogram of the test data. It should resemble the picture shown next.

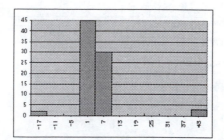

Open a text box and explain why this histogram presents a poor picture of the distribution of the data values.

(c) Draw a second histogram that eliminates the outliers and zooms in on the center of the previous histogram. An example is shown next.

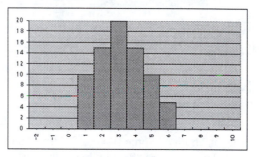

(d) Produce the histogram of the test data shown next.

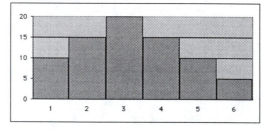

Which of the three histograms gives a better picture of the test data and why? Print the histograms with your comments in a text box.

Exercise 3.4 Open the worksheet entitled **test data**. In this exercise you will have the chance to compare two different boxplots drawn for this data set.
(a) Draw a boxplot treating all the data as one series, but otherwise use the default settings.
(b) Draw a second boxplot with the extreme percentile value set to 5%.
(c) Open a text box and discuss the two plots. Which gives a better picture of the data and why? Transfer the two boxplots to the worksheet containing the text box. Select the plots and text box and print them on a single page.

Drawing Parallel Boxplots

Exercise 3.5 Open the worksheet entitled **prof salary**. In this exercise you will draw boxplots of the salaries by rank and display these plots in the same chart. This chart will permit you to compare the distribution of salaries in the various ranks: Assistant Professor, Associate Professor, and Full Professor.

(a) Click in cell C1, then hold down the SHIFT key and click in cell E51.

(b) Launch Boxplot: pull down the Tools menu and release on Boxplot. In the Boxplot Parameters dialog box make sure that "Use first row/column as labels" and 0% are checked. Click the OK button. After resizing, your plot should look something like the next illustration. Print it.

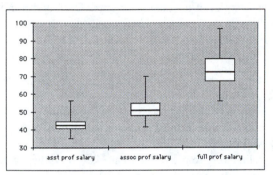

Exercise 3.6 These questions refer to the parallel boxplot drawn in the previous exercise. In (a) through (d) read the requested values from the plot and record them in a text box.

(a) The range of assistant professor salaries

(b) The median of associate professor salaries

(c) The 25th and 75th percentiles for full professor salaries

(d) The interquartile range for assistant professor salaries

(e) Open a text box and discuss the salary distributions. Compare their skewness. Which rank shows the greatest variability in salary? the least? How can you tell? Try to come up with a couple of explanations for the differences in variability.

(f) Which promotion seems to be the most lucrative: the one from assistant to associate professor or the one from associate to full professor? Again, how can you tell?

(g) Print the text boxes.

Exercise 3.7 Open the worksheet **lifespan**.

(a) Produce a boxplot of the lifespans listed in column B.

(b) Produce a histogram of the lifespans listed in column B. Experiment with the bin width until your histogram has a reasonably well-defined and detailed shape with no (or few) gaps.

(c) Describe the distribution of lifespans qualitatively (e.g. symmetrical, skewed to the right, double-peaked, etc.). Print a copy of the charts with your comments in a text box.

(d) Draw parallel boxplots of the male and female lifespan data. What does this data suggest about the relative lifespans of men and women? Explain your reasoning. Take into account the median values and the variability as shown by the the range and the interquartile range.

Exercise 3.8 Open the worksheet entitled **smoking habits** and read the accompanying text box that describes the data.
(a) Draw a histogram of the data in column C. What does it tell you about the smoking habits of most students?
(b) Draw another histogram of the same column-C data in which only the data for smokers is displayed.
(c) Draw a parallel boxplot of the data in columns D, E, and F in which the plots are labeled "men," "women," and "all." What does this chart tell you about the smoking habits of these students?
(d) Print the histograms and boxplots with accompanying text box discussions.

Exercise 3.9 Open the worksheet entitled **age guess gender** and read the accompanying text box that describes the data.
(a) Produce a histogram of the age guesses listed in column B in which the outliers have been eliminated. Experiment with the parameters until your histogram has a reasonably well-defined and detailed shape with no (or few) gaps. Describe the distribution qualitatively (e.g. symmetrical, skewed to the right, double-peaked, etc.). Also, explain and justify your criteria for determining outliers. Print a copy of the histogram with your comments in a text box.
(b) Produce a parallel boxplot of the male and female guesses. Which group was the better guesser? Explain your reasoning using the information provided by the plot. Print the plot and the discussion, organized nicely on the same page.

Exercise 3.10 Open the worksheet entitled **GPA data** and read the accompanying text box that describes the data.
(a) Produce a histogram of column B of **GPA data**. Redraw to eliminate the outliers. Experiment with the parameters until your histogram has a reasonably well-defined and detailed shape with no (or few) gaps. Describe the distribution qualitatively (e.g. symmetrical, skewed to the right, double-peaked, etc.). Also, explain and justify your criteria for determining outliers. Print a copy of the histogram with your comments in text boxes.
(b) Produce parallel boxplots of the GPAs sorted by college. Which college has the better high school GPA record? Explain your reasoning. Take into account the median values and the variability as shown by the the range and the interquartile range.

Exercise 3.11 Open the worksheet entitled **GPA gender** and read the accompanying text box that describes the data. Produce parallel boxplots of the GPAs sorted by gender. Which group has the better high school GPA record? Explain your reasoning. Take into account the median values and the variability as shown by the the range and the interquartile range. Print the plots and text box discussions.

Exercise 3.12 Open the worksheet entitled **height-gender** and read the accompanying text box that describes the data.

(a) Produce a histogram of all of the height data in which the outliers have been eliminated. Experiment with the histogram parameters until your histogram has a reasonably well-defined and detailed shape with no (or few) gaps. Describe the distribution qualitatively (e.g. symmetrical, skewed to the right, double-peaked, etc.). Also, explain and justify your criteria for determining outliers. Print a copy of the histogram with your comments in text boxes.

(b) Repeat part (a) for the male height data.

(c) Repeat part (a) for the female height data.

(d) In a text box explain how the histograms in parts (b) and (c) are related to the one in part (a). Print the charts and text box discussions.

Exercise 3.13 Open **datach2.xls** from **STAT** and turn to the worksheet **exercise-pulse**.

(a) Use IF (See Section 2.5) to sort the pulse rate data into exerciser and nonexerciser columns. Draw parallel boxplots of the data.

(b) What do the boxplots suggest about the relationship between pulse-rate and exercise?

WHAT TO HAND IN: Print the histograms and boxplots with commentary answering each question. Make sure each is labeled so it is clear which data set you have used.

Sorting and Describing Data

4

In This Chapter...

- AutoFilter
- Creating names
- Descriptive statistics
- The Rank and Percentile Tool

At the heart of statistics is the organization and analysis of data. Previous chapters have presented a variety of geometric and numerical techniques that distill raw chunks of data values into a few representative numbers or a chart. Principal among these were the measures of central tendency and spread (such as the mean and standard deviation) and the pictorial representation of data in boxplots and histograms. In this chapter several new Excel utilities are introduced that, in many instances, can streamline and expedite this process of organization and analysis. The tools used are contained in the Excel's Analysis ToolPak which must be added into your version of Excel. If you haven't already done so, see Appendix B for instructions.

To introduce these new features consider the following problem: a varsity third baseman with an ambition to play in the major leagues would like to know how much money a third baseman earns. Of course, different major league teams pay their third basemen different amounts, so there is no easy answer to the question. There is data—the salaries actually earned by third basemen—and this data must be analyzed by the varsity player if he is to get some sense of his financial prospects. The first step is to find relevant data. A little bit of Web surfing leads to a list of the salaries paid to major league players in 1994. This same data can be found in **datach4.xls**. Select the sheet labeled **all-bb**.[1] It will look like the illustration shown next.

---

[1]Copyright 1994, USA TODAY. Reprinted with permission.

	A	B	C	D	E	F	G	
1	Team	Player	Salary	Position				
2	Atlanta Braves	Fred McGriff	3750000	1b				
3	Atlanta Braves	Ryan Klesko	111500	2b				
4	Atlanta Braves	Mark Lemke	1100000	2b				
5	Atlanta Braves	Terry Pendleton	3200000	3b				
738	Toronto Blue Jays	Pat Hentgen	500000	p				
739	Toronto Blue Jays	Tony Castillo	437500	p				
740	Toronto Blue Jays	Woody Williams	150000	p				
741	Toronto Blue Jays	Scott Brow	115000	p				
742	Toronto Blue Jays	Paul Spoljaric	109000	p				
743	Toronto Blue Jays	Duane Ward	4000000	p				
744	Toronto Blue Jays	Danny Cox	800000	p				
745	Toronto Blue Jays	Paul Menhart	109000	p				
746	Toronto Blue Jays	Dick Schofield	640000	ss				
747	Toronto Blue Jays	Domingo Cedeno	122500	ss				
748	Toronto Blue Jays	Alex Gonzalez	109000	ss				
749								
750								

The 1994 salaries of all major league baseball players as reported in USA Today on April 5, 1994. Salaries include pro-rated signing bonuses.

Retrieved at: http://www.geom.umn.edu/docs/snell/chance /teaching_aids/baseball_salaries.html

all-bb / billionaires / heights / gpacol / Sheet8 / Sheet9

Each 1994 major-league player is in this list together with his team name, salary, and position. There are several hundred entries. The third basemen are indicated by the code "3b" in the Position column. We will begin by extracting the third basemen's salaries from the list. The first problem, that of sorting the third basemen's salary data from all the rest, can be quickly done with an Excel utility called *AutoFilter*.

4.1 USING AUTOFILTER

An Example

1. Split the screen so that rows 1 and 748 are both visible.
2. Select the contents of the worksheet—click in cell A1, hold down the SHIFT key and click in cell D748.
3. Pull down the Data menu and select Filter and then AutoFilter.

Drop-down buttons will appear to the right of each column label as shown next.

	A	B	C	D
1	Team ▼	Player ▼	Salary ▼	Position ▼
2	Atlanta Braves	Fred McGriff	3750000	1b
3	Atlanta Braves	Ryan Klesko	111500	2b
4	Atlanta Braves	Mark Lemke	1100000	2b
5	Atlanta Braves	Terry Pendleton	3200000	3b

When you click on one of these buttons, Excel will display a menu listing all of the different entries that appear in the column below the button. The menu for the Position column is shown next.

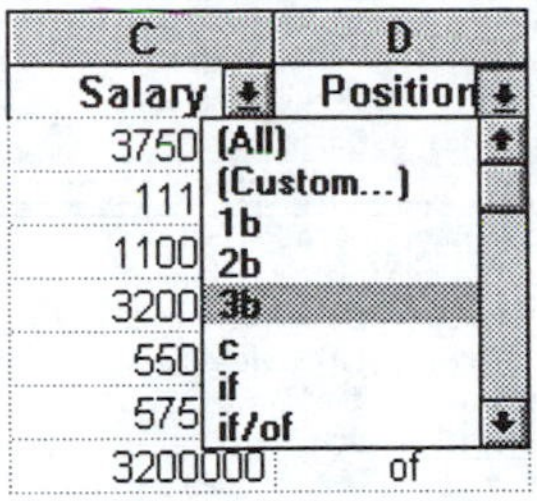

C	D
Salary	Position
3750	(All)
111	(Custom...)
	1b
1100	2b
3200	3b
550	c
	if
575	if/of
3200000	of

4. Select **3b** from the list of positions as shown above. Excel will extract all of the third basemen's data from the salary list. The result will look like the next picture.

	A	B	C	D
1	Team ▼	Player ▼	Salary ▼	Position ▼
5	Atlanta Braves	Terry Pendleton	3200000	3b
32	Baltimore Orioles	Chris Sabo	2000000	3b
58	Boston Red Sox	Sctt Cooper	475000	3b
110	Chicago Cubs	Steve Buechele	2550000	3b
457	New York Mets	Bobby Bonilla	6300000	3b
481	New York Yankees	Wade Boggs	3100000	3b
535	Philadelphia Phillies	Dave Hollins	2000000	3b
564	Pittsburgh Pirates	Jeff King	2400000	3b
589	San Diego Padres	Archi Cianfrocco	185000	3b
613	San Francisco Giants	Steve Scarsone	130000	3b
614	San Francisco Giants	John Pattersan	115000	3b
615	San Francisco Giants	Matt Williams	4050000	3b
640	Seattle Mariners	Edgar Martinez	3316667	3b
641	Seattle Mariners	Mike Blowers	300000	3b
668	St. Louis Cardinals	Todd Zeile	2700000	3b
669	St. Louis Cardinals	Stan Royer	130000	3b
696	Texas Rangers	Dean Palmer	475000	3b
724	Toronto Blue Jays	Damell Coles	500000	3b
725	Toronto Blue Jays	Ed Sprague.	500000	3b
749				
750				

Notice that the numbers of the selected rows have changed color. The other rows of the salary data, those for the positions other than third base, are not displayed. They have not been lost. In fact, if you select **(All)** from the drop-down menu, Excel will restore the complete list of salaries to the worksheet.

5. Copy the filtered third-baseman data to a new worksheet.
    (a) Click in A1, hold down the SHIFT key and click on the last salary value in column C.
    (b) Select Copy from the Edit menu.
    (c) Open a new workbook and click in cell A1.
    (d) Select Paste from the Edit menu.

	A	B	C	D
1	Team	Player	Salary	
2	Atlanta Braves	Terry Pendleton	3200000	
3	Baltimore Orioles	Chris Sabo	2000000	
4	Boston Red Sox	Sctt Cooper	475000	
27	San Francisco Giants	John Pattersan	115000	
28	San Francisco Giants	Matt Williams	4050000	
29	Seattle Mariners	Edgar Martinez	3316667	
30	Seattle Mariners	Mike Blowers	300000	
31	St. Louis Cardinals	Todd Zeile	2700000	
32	St. Louis Cardinals	Stan Royer	130000	
33	Texas Rangers	Dean Palmer	475000	
34	Toronto Blue Jays	Damell Coles	500000	
35	Toronto Blue Jays	Ed Sprague.	500000	
36				
37				
38				

sal3b / Sheet2 / Sheet3 / Sh

Notice that the screen was split, columns A and B were widened, and the new worksheet was named **sal3b**.

Turning AutoFilter Off

When you are done filtering a data set, remove the drop-down buttons from the worksheet containing the data. Since AutoFilter is a toggle, it can be turned off exactly as it was turned on.

1. Select the worksheet containing the data with the AutoFilter buttons.
2. Select Data from the menu bar and then Filter and AutoFilter. Excel will remove the check from AutoFilter in the menu and, at the same time, the drop-down buttons from the worksheet.

4.2 NAMES IN EXCEL

How to Name a Range of Data

Before beginning an analysis of the data in the **sal3b** worksheet, pause to learn a little trick that will make this task and similar ones easier to manage. The idea is to give the column of salary figures the name "Salary" so that it can be referred to without having to type its range address "C2:C35."

1. Select the range C1:C35. (Note that this range contains all of the salary data as well as the name, in row 1, to be assigned the data.)
2. Pull down the Insert menu and select Name and then Create... . A dialog box will open.

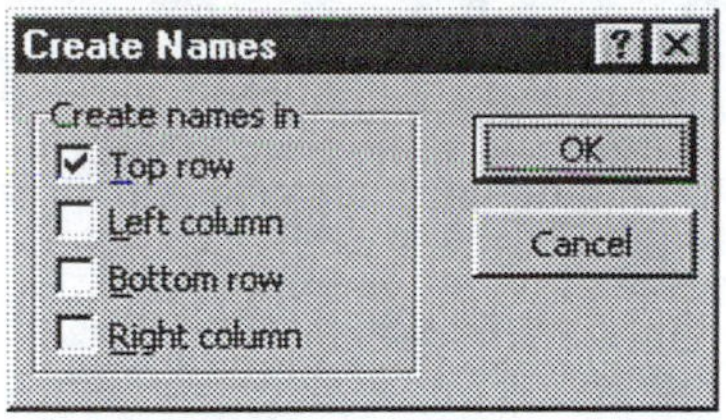

3. Click in the **Top Row** box to check it, as shown above, and then click the OK button.

Now, whenever you type the word "Salary" into a command, Excel will understand that you mean the column of salary data in this worksheet.

Using a Name in a Command

Use this newly minted name to find the average third baseman's salary. Simply click in the cell where you wish the average to be placed and type in the command =AVERAGE(Salary). Press the ENTER key. The result is shown in the illustration below in cell K2. Note that column K was widened and formatted as dollars.

	A	B	C	I	J	K	L
1	**Team**	**Player**	**Salary**				
2	Atlanta Braves	Terry Pendleton	3200000		**average**	$ 1,522,299	
3	Baltimore Orioles	Chris Sabo	2000000				
4	Boston Red Sox	Sctt Cooper	475000				
5	Chicago Cubs	Steve Buechele	2550000				
6	Chicago White Sox	Julio Franco	1000000				

K2 = =AVERAGE(Salary)

The Names Menu

The names created for a workbook can be viewed by pulling down the Names menu as shown below. To see the menu, click on the drop-down arrow located to the left of the formula bar.

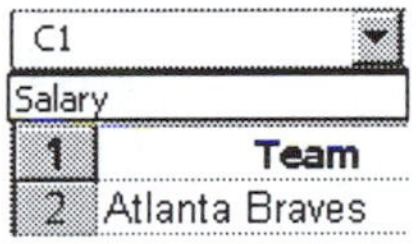

The Names menu list contains a single name, Salary, the only one created so far for this workbook.

Deleting a Name

A name can be removed from the list of defined names as follows.

1. Pull down the Insert menu and select Name and then Define... .

2. When the Define Name dialog box appears, select the name to be removed by clicking on it.
3. Click the Delete button and the selected name will be removed from the list.
4. Click the OK button.

4.3 DATA ANALYSIS: DESCRIPTIVE STATISTICS

The average is just one of many possible numerical measures of a data set. This section contains instructions for an Excel Data Analysis tool called *Descriptive Statistics*, which will return an entire table of important statistical measures.

1. Pull down the Tool menu and select Data Analysis... .
2. Select Descriptive Statistics from the Data Analysis dialog box as shown next.

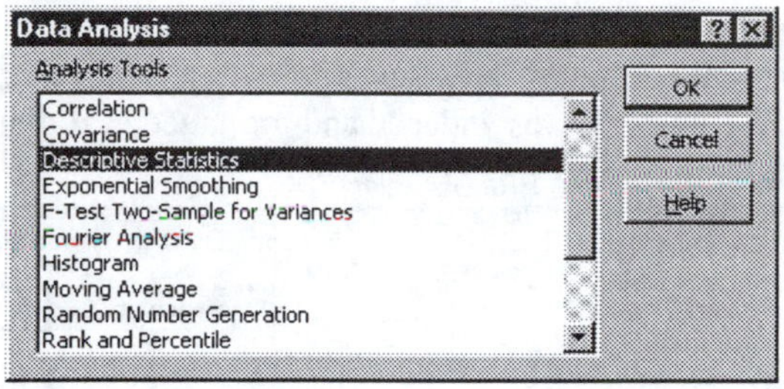

3. When the Descriptive Statistics dialog box opens, type "Salary" into the the **Input Range**. Make sure that **Group By Columns** is selected.
4. Click the **Output Range**. Click in the range box and type the cell address where the upper left corner of the table will be placed (in this case, E1).
5. Select **Summary Statistics**. The dialog box should look like the next illustration.

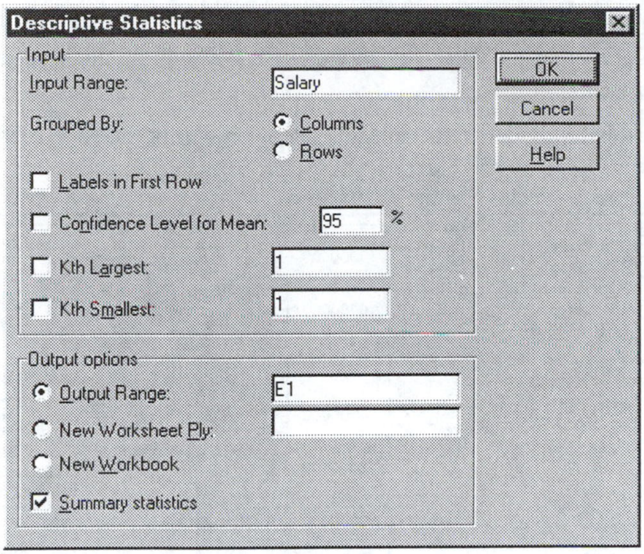

Descriptive Statistics

Input

Input Range: `Salary`

Grouped By: ⊙ Columns
             ○ Rows

☐ Labels in First Row

☐ Confidence Level for Mean: `95` %

☐ Kth Largest: `1`

☐ Kth Smallest: `1`

Output options

⊙ Output Range: `E1`

○ New Worksheet Ply:

○ New Workbook

☑ Summary statistics

[OK] [Cancel] [Help]

6. Click the OK button.

The Descriptive Statistics table for the salary data will be displayed in worksheet **sal3b** as shown below.

	A	B	C	D	E	F	G
1	Team	Player	Salary		*Column1*		
2	Atlanta Braves	Terry Pendleton	3200000				
3	Baltimore Orioles	Chris Sabo	2000000		Mean	1522299.03	
4	Boston Red Sox	Scott Cooper	475000		Standard Error	271245.082	
5	Chicago Cubs	Steve Buechele	2550000		Median	500000	
6	Chicago White Sox	Julio Franco	1000000		Mode	500000	
7	Chicago White Sox	Robin Ventura	3500000		Standard Deviation	1581617.03	
8	Cincinnati Reds	Tony Fernandez	500000		Sample Variance	2.5015E+12	
9	Cincinnati Reds	Lenny Harris	450000		Kurtosis	0.63670297	
10	Cincinnati Reds	Willie Greene	125000		Skewness	1.06229143	
11	Cleveland Indians	Jim Thome	325000		Range	6191000	
12	Colorado Rockies	Charlie Hayes	3050000		Minimum	109000	
13	Detroit Tigers	Scott Livingstone	365000		Maximum	6300000	
14	Houston Astros	Ken Caminiti	3200000		Sum	51758167	
15	Houston Astros	Chris Donnels	170000		Count	34	
16	Kansas City Royals	Gary Gaetti	109000		Confidence Level(95	551852.707	
17	Los Angeles Dodgers	Tim Wallach	3412500				
34	Toronto Blue Jays	Darnell Coles	500000				
35	Toronto Blue Jays	Ed Sprague.	500000				
36							

Don't worry if you are not familiar with every statistic given in this table. Some of them, namely, Standard Error, Sample Variance, and Confidence Level, are important in statistical inference, a topic that will be investigated later in this manual. For quick reference all of these statistics are briefly described at the end of this chapter.

Meanwhile, a perusal of the statistics that are familiar tells a clear tale of the distribution of the data. Notice that the maximum salary paid a third baseman is $6,300,000, a nice piece of change. However, the lowest-paid third baseman makes, in comparison, considerably less, a meager $109,000. Even more sobering is the median salary of $500,000. Half the third basemen earn less than this amount. Compare the median to the mean: the high average salary of \$1,522,299.03 is three times the median. These statistics paint a picture of a set of salary values that are skewed widely to the right by the salaries of a few very highly paid players.

A boxplot of this salary data tells the same tale. You will be asked to recreate this plot in the first exercise.

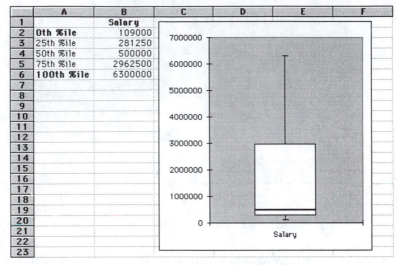

Clear the Descriptive Statistics table from the worksheet before going on, or move it out of the way, five or six columns to the right, to make room for a second Data Analysis table.

4.4 DATA ANALYSIS:
THE RANK AND PERCENTILE TABLE

Excel's Rank and Percentile table displays a data set in ranked order from highest to lowest. Apply this tool to the salary data.

How to Use the Tool

1. Open the worksheet that contains the data to be ranked, in this case **sal3b**.
2. Pull down the Tools menu and select Data Analysis... .
3. When the Data Analysis dialog box appears, scroll down and select Rank and Percentile as shown next.

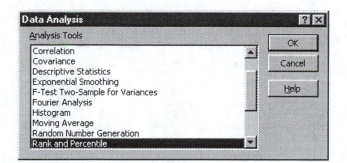

4. When the Rank and Percentile dialog box opens, type "Salary" into the **Input Range**. Click **Output Range** and type in the address of the cell where you wish the upper-left corner of the table placed. In the example below, it is $E$1, but you may use a different address if you like. The $ signs can be omitted.

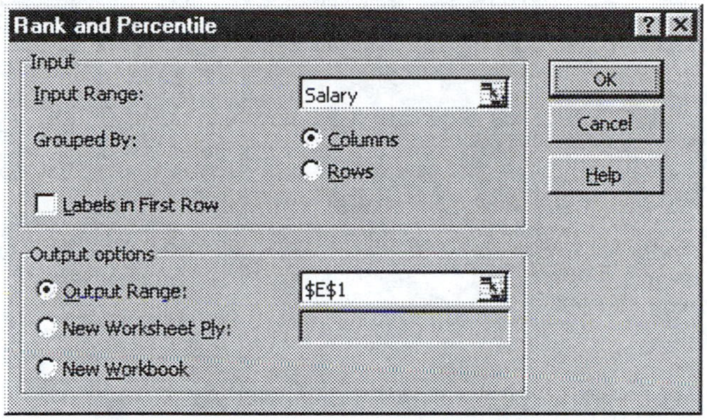

5. Excel will sort the salary data, placing it in descending order from the largest salary down to the smallest. It will also provide a percentile ranking for each salary value. The result will look something like the next illustration.

	A	B	C	D	E	F	G	H
	Team	Player	Salary		Point	Column1	Rank	Percent
2	Atlanta Braves	Terry Pendleton	3200000		20	6300000	1	100.00%
3	Baltimore Orioles	Chris Sabo	2000000		27	4050000	2	96.90%
4	Boston Red Sox	Sctt Cooper	475000		6	3500000	3	93.90%
5	Chicago Cubs	Steve Buechele	2550000		16	3412500	4	90.90%
6	Chicago White Sox	Julio Franco	1000000		28	3316667	5	87.80%
7	Chicago White Sox	Robin Ventura	3500000		1	3200000	6	81.80%
8	Cincinnati Reds	Tony Fernandez	500000		13	3200000	6	81.80%
9	Cincinnati Reds	Lenny Harris	450000		21	3100000	8	78.70%
10	Cincinnati Reds	Willie Greene	125000		11	3050000	9	75.70%
11	Cleveland Indians	Jim Thome	325000		30	2700000	10	72.70%
12	Colorado Rockies	Charlie Hayes	3050000		4	2550000	11	69.60%
13	Detroit Tigers	Scott Livingstone	365000		23	2400000	12	66.60%
14	Houston Astros	Ken Caminiti	3200000		2	2000000	13	60.60%
15	Houston Astros	Chris Donnels	170000		22	2000000	13	60.60%
16	Kansas City Royals	Gary Gaetti	109000		5	1000000	15	57.50%
17	Los Angeles Dodgers	Tim Wallach	3412500		18	650000	16	54.50%
18	Los Angeles Dodgers	Dave Hansen	275000		7	500000	17	45.40%
19	Milwaukee Brewers	Kevin Seitzer	650000		33	500000	17	45.40%
20	Montreal Expos	Sean Berry	200000		34	500000	17	45.40%
21	New York Mets	Bobby Bonilla	6300000		3	475000	20	39.30%
22	New York Yankees	Wade Boggs	3100000		32	475000	20	39.30%
23	Philadelphia Phillies	Dave Hollins	2000000		8	450000	22	36.30%
24	Pittsburgh Pirates	Jeff King	2400000		12	365000	23	33.30%
25	San Diego Padres	Archi Cianfrocco	185000		10	325000	24	30.30%
26	San Francisco Giants	Steve Scarsone	130000		29	300000	25	27.20%
27	San Francisco Giants	John Patterson	115000		17	275000	26	24.20%
28	San Francisco Giants	Matt Williams	4050000		19	200000	27	21.20%
29	Seattle Mariners	Edgar Martinez	3316667		24	185000	28	18.10%
30	Seattle Mariners	Mike Blowers	300000		14	170000	29	15.10%
31	St. Louis Cardinals	Todd Zeile	2700000		25	130000	30	9.00%
32	St. Louis Cardinals	Stan Royer	130000		31	130000	30	9.00%
33	Texas Rangers	Dean Palmer	475000		9	125000	32	6.00%
34	Toronto Blue Jays	Damell Coles	500000		26	115000	33	3.00%
35	Toronto Blue Jays	Ed Sprague.	500000		15	109000	34	.00%
36								

all-bb / Boxplot1 \ sal3b / billionaires / heights

Reading the Rank and Percentile Table

Look at the first entry in column E of the table. It tells you that the highest-ranked salary in 1994, $6,300,000, was earned by the 20th third baseman (or Point) in the original list. His name and team (Bobbie Bonilla of the New York Mets) can be found in row 21. Excel has assigned this highest salary a rank of 1 and a percentile ranking of 100%. The lowest salary, that of the 15th third baseman, Gary Gaetti of the Kansas City Royals, is $109,000. Excel assigns this salary a rank of 34 and a percentile ranking of 0.00%.

Ranks are assigned to all of the salary values in order and listed in the Rank column. Equal salaries are given equal ranking. For example, $2,000,000 appears in lines 14 and 15 of the table. If these had been different values, they would have been given the rankings 13 and 14, instead Excel assigned both a rank of 13 and then skipped rank 14.

Percentiles are determined according to the following scheme: Since there are 33 intervals between data points, Excel allots each interval a percentile step of 1/33, which is just slightly more than 3%. Consequently, the percentile values drop by 3% as the values descend the percentile column. (There is some variation, presumably due to round-off error.) Equal salary values are assigned equal percentiles: that of the lowest percentile that would have been assigned the group had the values been different. This method of assigning percentiles leads to some very annoying inconsistencies; for example, the median salary value of $500,000 is not given the rank of 50th percentile (50%) as it should be. However, even with these limitations, it is possible to read important information from a Rank and Percentile table. It is easy to see that the average salary of $1,522,299 corresponds to a percentile ranking in the high 50's, well above the median. It also makes clear that the lower half of the salaries lie in a tight range from about $100,000 to $500,000, whereas the top 50% lie in a much broader range from $500,000 to $6,300,000. So, how much should our hypothetical third baseman expect to make in the majors? A multimillion dollar contract is possible but certainly not as likely as a much more modest amount in mid-six digits or less.

4.5 THE DESCRIPTIVE STATISTICS TABLE

As promised, here is a brief description of the statistics listed in this table.

- The **Mean** is the average of the data values.
- The **Standard Error** is the standard deviation of the data values divided by the square root of Count.
- The **Median** is the middle data value.
- The **Mode** is the most frequently occuring data value. Not every data set has a modal value and when this is the case Excel returns **#N/A**.
- The **Standard Deviation** is the square root of Sample Variance.
- The **Sample Variance** is the sum of the squared deviations from the Mean divided by Count minus 1.
- **Kurtosis** is a measure of the extent to which the distribution of the data has a central peak when compared to the normal distribution. A negative kurtosis indicates a data distribution that is less peaked than a normal distribution, and a positive kurtosis one that is more peaked.
- **Skewness** measures the degree of symmetry about the mean. A positive skewness value indicates a distribution that is skewed to the right, in other words, that has a long asymmetric tail extending to the right of the mean. Negative skewness indicates data that is skewed to the left.
- The **Range** is the difference between the maximum and the minimum data values.
- **Minimum** refers to the smallest data value and **Maximum** to the largest.
- **Sum** is the total of all of the data values.
- **Count** is the number of data values.
- The default **Confidence Level** gives the half-width of the 95% confidence interval for the population mean with the data values regarded as a sample from this population. (Depending on what version of Excel you are using, this statistic may not be

automatically included in the Summary Statistics option. If not, it can be added by selecting the Confidence Interval option from the Descriptive Statistics dialog box.)

EXERCISES

The data sets for these exercises are in the workbook **datach4.xls**.

Exercise 4.1 Produce the Descriptive Statistics table and the Rank and Percentile table for the third-baseman salary data as described in Section 4.3 and Section 4.4. Draw a boxplot of the data. Print them.

Exercise 4.2 Open the worksheet called **heights**.
(a) Use AutoFilter to extract the female heights. Copy these heights to a new workbook.
(b) Create the name "Height" for the list of female heights.
(c) Generate a Descriptive Statistics table for the height data and a boxplot.
(d) Open a text box and, referring to the table and the plot, write a description of the data distribution.
(e) Print the table and the description. Do not print the data.

Exercise 4.3 Produce a Rank and Percentile table and a Descriptive Statistics table for the salaries of major league pitchers.
(a) Who was the highest-paid pitcher in 1994?
(b) Find the name of a pitcher whose salary was close to the median.
(c) How many pitchers were paid the lowest salary?
(d) Why is the mode a poor measure of central tendency for this data set?
(e) Describe the salary prospects of a major league pitcher based on the 1994 data.

Exercise 4.4 Produce a Rank and Percentile table for the salaries of major league catchers.
(a) Who was the highest-paid catcher in 1994?
(b) Who was the lowest-paid catcher?
(c) Find the name of a catcher whose salary was close to the median.
(d) Draw a boxplot of the data and write a description of the distribution that refers to table and the boxplot.

Exercise 4.5 Open the **billionaires** worksheet. Contruct a Descriptive Statistics table for the age of Middle Eastern billionaires. Is this distribution skewed? Explain how you can tell from the information in the table.

Exercise 4.6 Open the **books** worksheet.
(a) Construct a Rank and Percentile table for the pleasure-reading of female Arts College majors.
(b) Construct a Descriptive Statistics table for the same group.
(c) Construct a boxplot of the data.
(d) Open a text box and write a discussion of the distribution using the information contained in the tables and relating it to the shape of the plot.

Exercise 4.7 Follow the instructions in the previous problem for male Business College majors.

WHAT TO HAND IN: Print the Excel worksheet displaying the calculations, tables, charts and text boxes required for each problem assigned.

The Chart Wizard

In This Chapter...

- Chart Wizard vocabulary
- Using Chart Wizard
- Resizing and moving charts in a worksheet

The boxplots and histograms studied in Chapter 3 are powerful techniques of visual summary. Each can transform a disorganized chunk of raw data into a picture that reveals the basic character of the numbers: their center, their shape, and their spread. In this chapter we will look at a different set of charting techniques. These work primarily on data that has already been organized into table form, so the purpose of the chart is to reveal relationships among the various table entries such as relative size or changing magnitude over time. A well constructed chart or graph of this sort can highlight features of the data that might otherwise be difficult to see. Charts and graphs can be easily produced in Excel using a special utility called **Chart Wizard**.

The Chart Wizard utility in Excel 97 works differently from the versions contained in Excel 5 and 7. The illustrations in this chapter correspond to the Chart Wizard dialog boxes for Excel 5 and 7. The Excel 97 version of Chart Wizard is described in Appendix C.

Begin by studying the following table of teen crime data.[1]

---

[1]SOURCE: *Sourcebook of Bureau of Justice Statistics*, U.S. Department of Justice, 1982 and 1991

Teen Arrests for Serious Crime (in thousands)		
	1980	1990
Violent Crime	66.4	68.7
Robbery	34.8	21.5
Burglary	145	150
Larceny	295	303
Auto Theft	41.2	34
Arson	5.2	6

5.1 THE LANGUAGE OF CHARTS

Chart Wizard was used to produce the following column chart from the Excel worksheet displayed after it.

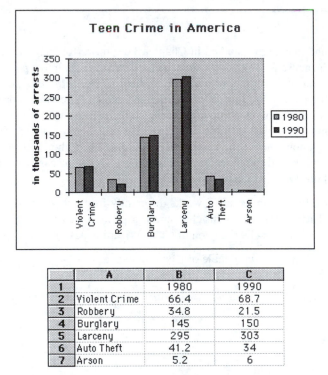

	A	B	C
1		1980	1990
2	Violent Crime	66.4	68.7
3	Robbery	34.8	21.5
4	Burglary	145	150
5	Larceny	295	303
6	Auto Theft	41.2	34
7	Arson	5.2	6

In order to produce such charts confidently, you need to know the language of charts. If you understand what the parts of the chart are called, then you will find it easier to produce the output that you want. If you don't understand this language, you will be at the mercy of Chart Wizard.

The data in the worksheet above is organized into three *data series*: a list of crimes, data for 1980, and data for 1990. Each data series is in a worksheet column. In producing the chart the first data series, the crimes, was used as titles for the horizontal axis (also called the x-axis). The 1980 data series and the 1990 data series became chart columns, with different-colored bars for each series. Keep in mind that the phrase "chart columns" refers to the bars of the chart itself and not to the worksheet columns where the original data was typed. At the right of the chart is a box, showing which color corresponds to which data series. This box is called a *legend*. Notice that the chart title, "Teen Crime in America," and the vertical or y-axis title, "in thousands of arrests," do not appear in the worksheet. In the next section, you will learn how to attach such labels to a chart. With this vocabulary in mind, let's look at how Chart Wizard works.

5.2 USING CHART WIZARD

Getting Ready

1. Open an Excel workbook and copy the teen crime data from **crime stats** in **datach5.xls** into the new workbook.

2. Highlight the data—click in cell A1 and drag to cell C7. The result should look something like this:

	A	B	C
1		1980	1990
2	Violent Crime	66.4	68.7
3	Robbery	34.8	21.5
4	Burglary	145	150
5	Larceny	295	303
6	Auto Theft	41.2	34
7	Arson	5.2	6

Make a mental note that the highlighted range is $A$1:$C$7. It will be referred to again.

3. Click on the Chart Wizard Button. It is pictured next. Look for it on the Standard Toolbar.

The mouse pointer will become the Chart Wizard cursor shown next.

The words "Drag in document to create a chart" appear at the bottom left of the screen. This instruction is a prompt asking you to outline the rectangular area of the worksheet where your chart will eventually be placed when it is complete.

4. Click at the point in the worksheet where you want the upper left-hand corner of the chart located and drag diagonally to the lower right-hand corner of the chart's eventual position. An outline of the range you have indicated will appear on the worksheet. An example is shown below. Don't worry if your region has a different size and location.

	A	B	C	D	E	F
1		1980	1990			
2	Violent Crime	66.4	68.7			
3	Robbery	34.8	21.5			
4	Burglary	145	150			
5	Larceny	295	303			
6	Auto Theft	41.2	34			
7	Arson	5.2	6			
8						
9						
10						

When you release the mouse, a dialog box will open and the position you outlined will disappear. This happens very quickly. Don't be concerned: Excel will remember the position you indicated and place your chart there once it is complete. You will then have an opportunity to relocate the chart on the worksheet and, if you like, to change its size.

Step 1 of Chart Wizard

The following dialog box is the first of five Chart Wizard steps. Its purpose is to make sure that Excel understands where your data is located. The range indicated in the Range box, $A$1:$C$7, is correct. If it were not, you would type in the correct range now before going on.

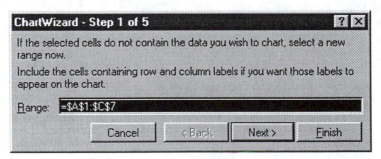

Since the indicated range *does* contain the data you want to chart, click the Next button.

Step 2 of Chart Wizard

The Chart Type dialog box will appear. It consists of an array of chart options.

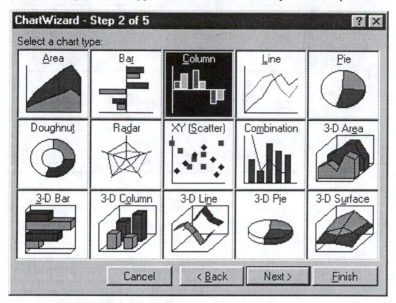

You select the type you prefer by clicking on the square containing it. However, in this case, the column option is already highlighted. It is the *default* choice, which means that if you make no selection, Excel will automatically draw a column chart. Since this is indeed the type of chart you want, all you have to do at Step 2 is click the Next button.

Step 3 of Chart Wizard

Step 3 presents the Chart Format dialog box. It is also an array, this time of column-chart options. Format 6, the default option, is *not* the one you want, since it produces a bar chart with a background of grid lines. Format 1 omits the gridlines, so click on it and then on the Next button.

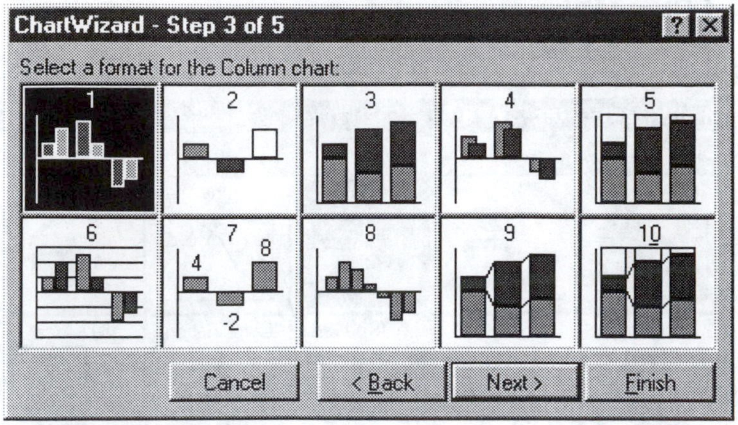

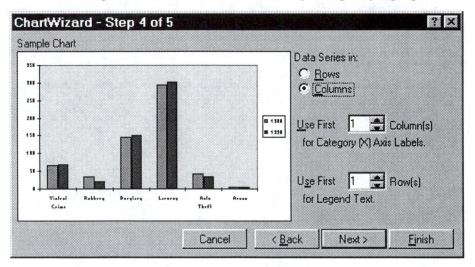

Quick Tip: Use the TAB key to move from line to line in any dialog box. Do not press the ENTER key unless you are ready to go on to the next Chart Wizard step.

Step 4 of Chart Wizard

The next dialog box, Step 4, gives a rough preview of what the chart will look like and provides a set of buttons for altering its appearance. The questions in the right-hand column of the dialog box refer to the worksheet data that you originally highlighted.

In the Step 4 dialog box, Chart Wizard asks for three answers.

1. *Data Series in Rows or Columns?* In the original worksheet, the data series for 1980 and 1990 are in columns. That is the way we want the data organized, so the Columns option must be selected. Remember, the word "columns" here refers to the columns of data in the worksheet, not the colored columns displayed in the sample chart. As an experiment try selecting the Rows option. The chart changes drastically, as the following illustration shows.

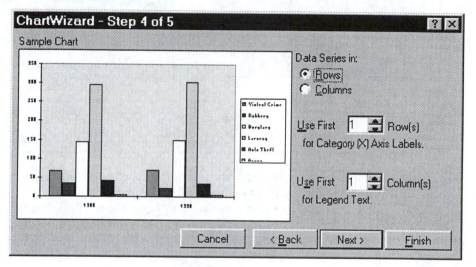

Reselect the Columns option now before going on.

2. *Use First Column for Category (X) Axis Labels?* Excel wants to know where it should look for labels to place on the horizontal or x-axis. In the original worksheet, the first column contains the names of the crimes, so leave this option set at the number 1.

3. *Use First Row for Legend Text?* The first row of the data contains the years 1980 and 1990. Certainly we do not want them graphically displayed; we want them to appear as they do—as legend text. So, leave this setting as it is.

When you have responded to all of the questions asked by the Step 4 dialog box, click the Next button.

Step 5 of Chart Wizard

The last dialog box, Step 5, permits you to add or delete a legend and to title your chart and its axes.

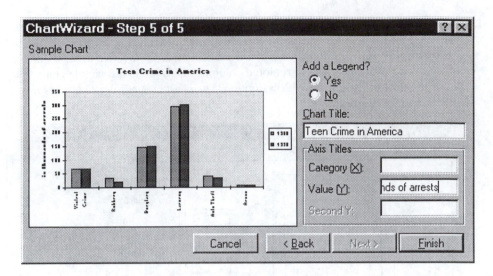

The chart needs a legend, so the answer to the first question in the dialog box should remain "Yes." In the future, if you don't want to see a legend, click the No button. Click in the **Chart Title** box and type "Teen Crime in America." In the **Value(Y)** title box type the label "in thousands of arrests." Leave the **Category(X)** title box blank. Click the Finish button and the chart will be embedded in your worksheet in the position you originally outlined in Step 3 above. Your worksheet will look something like the next illustration.

	A	B	C	D	E	F
1		1980	1990			
2	Violent Crime	66.4	68.7			
3	Robbery	34.8	21.5			
4	Burglary	145	150			
5	Larceny	295	303			
6	Auto Theft	41.2	34			
7	Arson	5.2	6			
8						
9						
10						
11						
12						

5.3 SOME SIMPLE EDITING TECHNIQUES

You are now ready to put the finishing touches on the chart. You may need to make it larger, which is certainly the case in the previous illustration, or, perhaps, you would like to move it to another spot in the worksheet. Maybe you made some error along the way and the picture that Chart Wizard produced is completely wrong—say, a pie chart instead of a bar chart. In this case you need to delete the current chart and start over. Each of these actions—resizing, repositioning, and deleting—is an editing technique and can be done quite easily once the chart is selected. If you are already comfortable with these editing techniques, which were briefly covered in the previous chapter on boxplots and histograms, skip to the next section.

How to Select a Chart

A chart is *selected* when its border contains eight rectangular handles. Notice that the chart is selected in the previous illustration. Chart Wizard automatically selects any chart as it embeds it in a worksheet. If your chart does not show the handles, click inside the chart ONCE and the handles will appear. If you mistakenly click more than once, your chart will be *activated* rather than simply selected. An activated chart has a thick border (on some monitors it is grey in color) and also has handles. To deactivate an activated chart and return it to selected status, click once outside the chart border. To deactivate *and* deselect, click twice outside the border. We will have a lot more to say about the editing of activated charts in the next chapter.

How to Change the Size and the Position of a Chart

To resize the chart, select it by clicking once inside the chart. Then place the point of the arrow on one of the eight rectangular handles located on the charts border and click and drag the mouse. The chart will change size and shape depending on the direction of the drag. Experiment with the resizing until you get an idea of how it works.

To reposition a chart, make sure that it is selected. If no handles are visible, click on it once. When the chart is selected, place the point of the cursor anywhere on the chart itself, click, and drag the chart to a new position in the worksheet.

See if you can make your chart look like the one shown next.

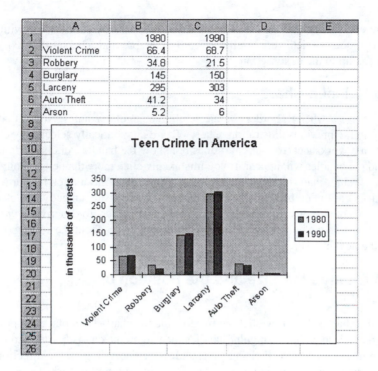

	A	B	C	D	E
1		1980	1990		
2	Violent Crime	66.4	68.7		
3	Robbery	34.8	21.5		
4	Burglary	145	150		
5	Larceny	295	303		
6	Auto Theft	41.2	34		
7	Arson	5.2	6		

How to Delete a Chart from a Worksheet

To remove a chart from a worksheet, first click on it once to select it and then press the DELETE key.

EXERCISES

The data for these exercises can be found in **datach5.xls**.

Exercise 5.1 Use Chart Wizard to produce the teen crime chart in the previous illustration. The data is located in the worksheet entitled **crime stats**. Select the chart and table and print them.

Exercise 5.2 A criminologist interested in teenage crime might like to see a chart of the percent change in the number of arrests for the various categories of crime over the ten-year period 1980–1990. Work through the steps below to produce such a chart.
 (a) Open the **crime stats** worksheet and type the phrase "Percent change" into cell D1.
 (b) Enter the formula =(C2-B2)/B2 into cell D2. Click the % key.
 (c) Now, move the cursor to the lower right corner of cell D2. When it becomes a black cross, double-click.
 (d) Select the range A2:A7 in the usual way. Now, hold down the CTRL key (the Apple key on a Macintosh) and highlight the range D2:D7. The worksheet should look like

the one shown next.

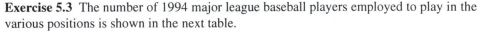

	A	B	C	D
1		1980	1990	Percent change
2	Violent Crime	66.4	68.7	3%
3	Robbery	34.8	21.5	-38%
4	Burglary	145	150	3%
5	Larceny	295	303	3%
6	Auto Theft	41.2	34	-17%
7	Arson	5.2	6	15%

(e) Produce the following chart of the teen crime data and print it.

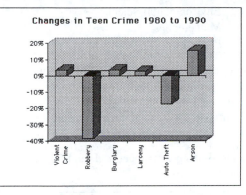

Exercise 5.3 The number of 1994 major league baseball players employed to play in the various positions is shown in the next table.

position	count
baseman	137
catcher	60
outfield	159
pitcher	331
shortstop	59

Display this data in a pie chart like the following one. It may be necessary to resize the chart in order to show all of the labels clearly. Print the chart.

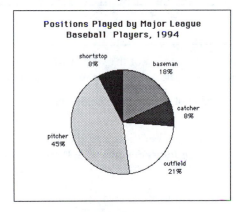

Exercise 5.4 The distribution of students throughout the major programs in a College of Science is shown below.

Distribution of 853 Science Students by Major	
Major	Percentage
Biology	53%
Chemistry	26%
Computer Science	21%
Environmental Science	11%
Mathematics	9%
Physics	5%

(a) Notice that the total percentage exceeds 100%. Is this necessarily due to an error in calculation? Can you think of a reason why it might be correct?

(b) Why is a pie chart inappropriate for this data?

(c) Design a chart that displays this data in a clear and accurate way.

(d) Print a text box containing your answers to (a) and (b) and the chart.

Exercise 5.5 Open the **skating** worksheet in **datach5.xls**.[2]

(a) Chart the data as shown next.

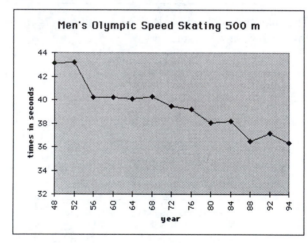

(b) In a text box write out a description of the way in which these times have changed over the years. Print the worksheet containing the chart and the discussion.

Exercise 5.6 Consider the table of capital punishment data[3] shown next.

---

[2]SOURCE: *The World Almanac and Book of Facts*, 1996, pg. 852.

[3]SOURCE: *Sourcebook of Bureau of Justice Statistics*, U.S. Department of Justice, 1994

Executions in the United States Since 1982

Year	Texas	In All States
1982	1	2
1983	0	5
1984	3	21
1985	6	18
1986	10	18
1987	6	25
1988	3	11
1989	4	16
1990	4	23
1991	5	14
1992	12	31
1993	14	30

Use this data, which you can find in worksheet **executions**, to produce the chart displayed below.

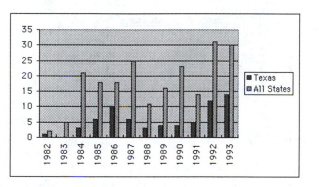

Exercise 5.7 Follow these steps to produce a bar chart comparing the number of Texas executions to the total number of executions in the other 49 states.

(a) Type the heading "Other States" into cell D2 and give the cell a lower border.

(b) Enter the command =(C3-B3) into cell D3 as shown next.

	A	B	C	D	E
1	*Executions*				
2	Year	Texas	All States	Other States	
3	1982	1	2	=(C3-B3)	
4	1983	0	5		
5	1984	3	21		
6	1985	6	18		
7	1986	10	18		
8	1987	6	25		
9	1988	3	11		
10	1989	4	16		
11	1990	4	23		
12	1991	5	14		
13	1992	12	31		
14	1993	14	30		

(c) Copy this command down column D by moving the cursor to the lower right corner of cell D3. When it becomes a black cross, double-click. Excel will compute the differences for all the years.

	A	B	C	D	E
1	*Executions*				
2	Year	Texas	All States	Other States	
3	1982	1	2	1	
4	1983	0	5	5	
5	1984	3	21	18	
6	1985	6	18	12	
7	1986	10	18	8	
8	1987	6	25	19	
9	1988	3	11	8	
10	1989	4	16	12	
11	1990	4	23	19	
12	1991	5	14	9	
13	1992	12	31	19	
14	1993	14	30	16	

(d) Now select the disconnected range A2:B14 and D2:D14 as displayed next. If you have forgotten how to make such a selection see Section 2.2.

	A	B	C	D	E
1	*Executions*				
2	Year	Texas	All States	Other States	
3	1982	1	2	1	
4	1983	0	5	5	
5	1984	3	21	18	
6	1985	6	18	12	
7	1986	10	18	8	
8	1987	6	25	19	
9	1988	3	11	8	
10	1989	4	16	12	
11	1990	4	23	19	
12	1991	5	14	9	
13	1992	12	31	19	
14	1993	14	30	16	

(e) Launch Chart Wizard and make the choices that produce the chart illustrated next.

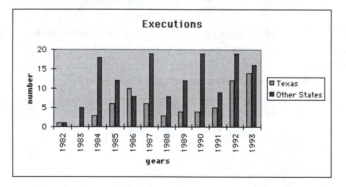

(f) Produce a two-color pie chart that compares the total number of Texas executions since 1982 to those in all the remaining states.

(g) Produce a chart that shows, for each year, the percentage of all executions that took place in Texas.

(h) Open a text box and write a discussion of the following proposition: Compared to other states, Texas has a disproportionate number of executions. What other data would be helpful in deciding this question?

(i) Print the charts and the text box nicely organized on a single sheet.

Exercise 5.8 Assume that you are keeping track of how many hamburgers and how many salads you eat each year, and your data set is entered into Excel as illustrated next.

	A	B	C
1	year	hamburgers	salads
2	88	40	5
3	89	43	10
4	90	38	29
5	91	22	63

This is, of course, made-up data, but this is OK because the purpose of the exercise is to practice with the various options available to you at step 4 of Chart Wizard. A small change in step 4 makes a big difference in the the appearance of the resulting chart and how it is interpreted.

Each of the following charts can be produced by Chart Wizard from the data set in the worksheet **food** after selecting the one range A1:C5. Test your knowledge of step 4 of the Chart Wizard by matching the following charts with the correct step 4 options. Record the step 4 settings for each in a text box and print the box and the charts.

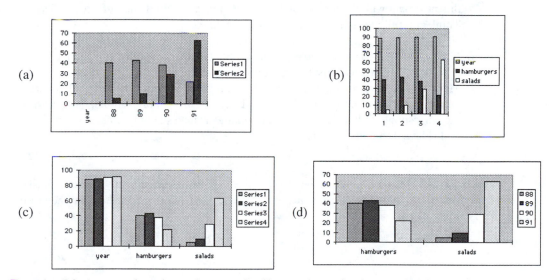

Exercise 5.9 A runner is trying to keep track of her workouts for the month prior to a big race. She entered the following data (numbers are in miles; no entry means she did not run that day) which can be found in the worksheet **workout times**.

	A	B	C	D	E
1		week 1	week 2	week 3	week 4
2	Monday	5	4	6	4
3	Tuesday	3	4	6	5
4	Wednesday		1	2	5
5	Thursday	3	5		
6	Friday	2			1
7	Saturday		2	5	7
8	Sunday	8	8	11	12

(a) Produce a column chart of this data set, in which the horizontal axis labels are "week 1", "week 2", etc. Above each week label there will be different-colored columns, one color for each day that she runs. For example, above the week 1 label there will be a cluster of five different-color columns, the first one 5 units high, the second 3 units high, etc. Make sure to include a legend that says what each color means, and include appropriate titles.

(b) Now produce a different style of column chart, with the same horizontal axis as before, but with only 4 columns, one for each week. Each column will have several different-color bands. For example, the week 1 column will be 21 units high, with the first level 5 units high, the second level 3 units high, etc. With this chart, unlike the first, you should be able to see the total weekly mileage at a glance, while you are still able to estimate how many miles she ran each day.

The two charts in parts (a) and (b) required only Chart Wizard. For these next two, you will first have to produce some new data (involving sums). Then you will select this new data with the mouse, and produce the chart with the aid of Chart Wizard. Think carefully what new data you need to calculate!

(c) Make a column chart that looks like the previous one except that it is monochrome (one color). In other words, this chart only shows the total weekly mileage.

(d) Make a pie chart showing the percent of the total mileage for the month that is run during each day of the week. The pie will have 7 colors. For example, the slice of the pie corresponding to Friday will be the smallest, since she only runs a total of 3 miles on Fridays during the month.

WHAT TO HAND IN: Print the charts and text boxes containing any discussion asked for. Make sure that the charts and text boxes are resized and positioned so that they fit on just a few pages.

6

More Chart Wizard

In Chapter 5 you learned how to use Chart Wizard to picture data tables. In this chapter you will continue to explore Excel's charting utility. Take a look at the following chart. It was drawn by the Wizard from data included in **datach6.xls** in the worksheet titled **crime**.[1]

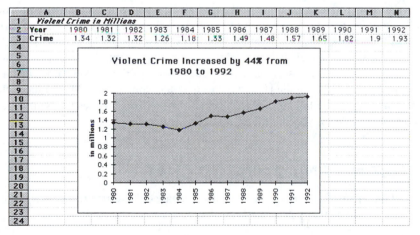

---

[1]SOURCE: FBI, *Uniform Crime Reports*, 1993.

This chart gives a more or less neutral and straightforward presentation of the data. The claim made in the title accurately reflects the facts. Now take a look at the two charts shown next. They were both drawn in Excel from exactly the same crime data. But as you can see by comparing them with the preceding one and with each other, the psychological impacts are quite different.

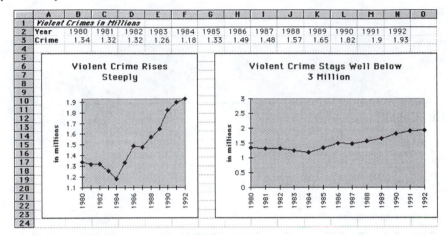

What produces the different effects? The titles, of course, make a difference. We are encouraged in these examples to view the violent crime increases in an extreme fashion, as very large in the left-hand chart and as negligible in the right-hand chart. In addition, the scalings on the vertical axes were adjusted and the charts reshaped to make the pictures seem to support these quite different views of exactly the same data set.

Pictures *can* exaggerate or even lie. In the next section you will learn how to produce charts like the two above. The point of the exercise is this: It is easy to produce charts that misrepresent data, and it is important that you know how this might be done so you can avoid being misled by a bad chart. In addition, there is a deeper problem with all three charts that goes beyond axis scaling or chart titles: None of them takes into account the changes in the population of the United States over the same period. Maybe the increase in crime was simply due to an increase in population. In the exercises you will have the chance to investigate for yourself the effect of this underlying variable.

6.1 EDITING AN ACTIVATED CHART

Excel contains many features that you can use to change the structure and format of an existing chart. You begin by *activating* the chart that you wish to edit.

How to Activate a Chart

Double-click the mouse inside the borders of the chart. When the chart is activated, the border changes its appearance and becomes, on most monitors, a wide, hatched frame around the chart with small solid square handles.

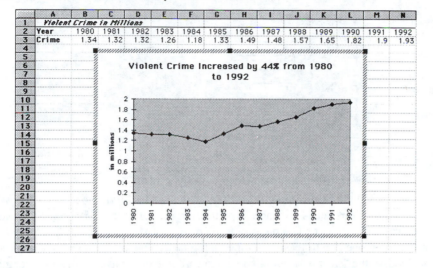

If your monitor is one that does not show this frame, you may see no border at all when the chart is activated. Only the square handles at the corners and on the edges will be visible.

A chart is deactivated by clicking in the worksheet outside the chart boundary.

How to Rescale a Chart Axis

In the chart shown on the first page of the chapter, the vertical axis is calibrated from 0 to 2. Suppose you wish to change this scale to the one shown on the chart titled "Violent Crime Rises Steeply," which runs from 1.1 to 1.93.

1. Activate the original chart.
2. Double-click on the vertical axis. Rectangles will appear at the top and bottom of the axis and the Format Axis dialog box will appear. (Or, click on the vertical axis once, pull down the Format menu and release on Selected Axis... .)
3. Since you want to scale the selected axis, click on the Scale tab.

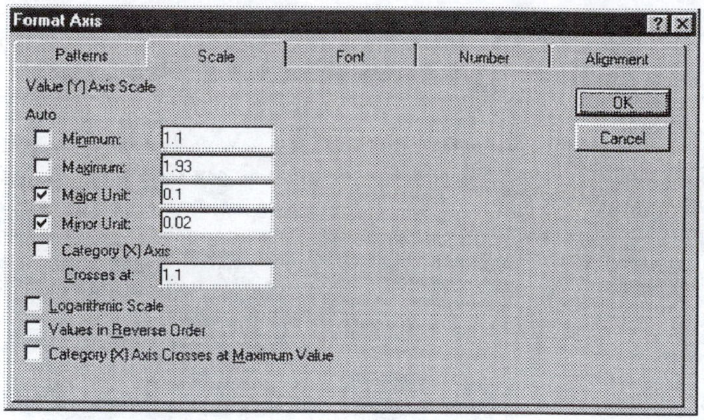

Format Axis dialog box:

Format Axis   ? ☒

| Patterns | Scale | Font | Number | Alignment |

Value (Y) Axis Scale

Auto

☐ Mi<u>n</u>imum: 1.1

☐ Ma<u>x</u>imum: 1.93

☑ Ma<u>j</u>or Unit: 0.1

☑ Mi<u>n</u>or Unit: 0.02

☐ <u>C</u>ategory (X) Axis

  C<u>r</u>osses at: 1.1

☐ <u>L</u>ogarithmic Scale
☐ Values in <u>R</u>everse Order
☐ Category (X) Axis Crosses at <u>M</u>aximum Value

OK
Cancel

4. Type "1.1" in the **Minimum** box and "1.93" in the **Maximum** box. Leave the **Major** and **Minor Units** set as they are. Type the number 1.1 in the **Category (X) Axis Crosses** box. Click the OK button. (Remember that you can use the TAB key to move from one line to the next in the dialog box.) You should now have a chart scaled like the "steeply rises" chart shown earlier. Note: you can adjust the length of the horizontal axis, a feature that differs dramatically in the three charts, by simply resizing the chart.

How to Edit a Chart Title

1. Activate the chart.
2. Click on the chart's title once or twice slowly (don't double-click) until the insertion point appears in the text of the title. Highlight the title you wish to change by clicking at the beginning of it and dragging to the end.

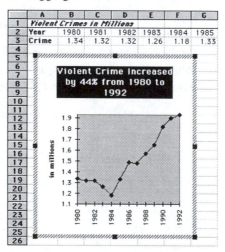

3. Type in the new title.

4. If you are done, double-click outside the chart to deactivate and deselect it.

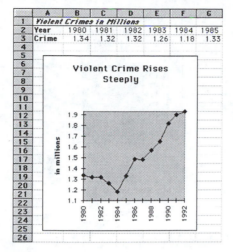

How to Add a Title to a Chart

Notice that in the previous chart the horizontal axis is not named. Suppose you decide to add the title "years" to this axis.

1. Activate the chart.

2. Pull down the Insert menu and select Titles... .

3. When the Titles dialog box appears, check **Category (X) Axis**.

4. Click the OK button.

5. A text region will appear at the bottom of the horizontal axis. In the next picture, it lies under 1986.

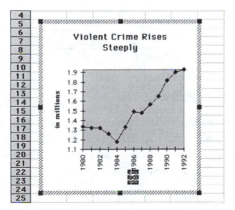

6. Type "years." It is not necessary to click in the text region, just type.

7. Deselect the chart, and it should now look like the one shown next.

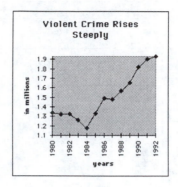

6.2 USING THE CHART TOOLBAR

The Chart Toolbar is a handy device that permits you to make certain changes in the appearance of a chart very quickly. This is not an essential feature, since all of the editing techniques on the Chart Toolbar can be done without the toolbar; however, the bar is a convenience, so you might want to spend a few minutes finding out how it works. If not, go on to Section 6.3: Adding Data to a Chart.

Viewing the Toolbar

Pull down the View menu and release on Toolbars... . (If the View menu is grayed out, click anywhere in the worksheet and then pull down the View menu.) When the Toolbox dialog box opens, click on Chart and then on OK. The toolbar shown below will appear on your worksheet.

The Chart Toolbar Buttons

1. The first button on the left allows you to change chart type. For example, if you wish to transform a line chart into a horizontal bar chart, you begin by activating the line chart. Next, click on the down arrow in the Chart Type button and select the horizontal bar chart option as shown below.

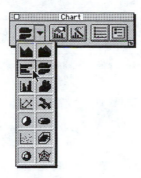

The activated line chart will become a horizontal bar chart.

2. The second button from the left on the Chart Toolbar will automatically change any chart to the default chart, which, unless the default has been reset in your version of Excel, is a vertical bar chart.

3. The third button launches Chart Wizard.

4. The fourth button is a toggle that places gridlines on an activated chart or removes them if they are already in place.

5. The fifth button is also a toggle. It can be used to add or delete a legend.

Experiment a little with the buttons of the Chart Toolbar.

6.3 ADDING DATA TO A CHART

Suppose you come across the violent crime data for 1993 (1,924,190) and want to add this information to the chart.

1. Open the workbook that contains the chart you wish to edit and the worksheet containing the chart data. Type the new data into the worksheet. In the next illustration the data was typed into cells O2 and O3 of the **crime** worksheet.

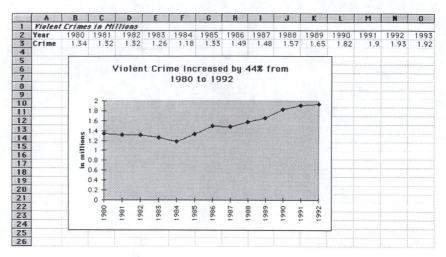

	A	B	C	D	E	F	G	H	I	J	K	L	M	N	O
1	*Violent Crimes in Millions*														
2	Year	1980	1981	1982	1983	1984	1985	1986	1987	1988	1989	1990	1991	1992	1993
3	Crime	1.34	1.32	1.32	1.26	1.18	1.33	1.49	1.48	1.57	1.65	1.82	1.9	1.93	1.92
4															

2. Activate the chart by double-clicking inside its border.

3. Pull down the Insert menu and release on New Data.... The New Data dialog box will appear.

4. Click once outside the chart. Then click and drag to select the range containing the new data, in this case, O2:O3. Excel will automatically write this range into the dialog box. Don't be concerned about the sheet name; yours may be different. Press the OK button.

5. The Paste Special dialog box opens next. In it Excel asks several questions about the nature of the data you are adding.

As the data will create an additional point on the chart, leave the **Add Cells as** option set to **New Point(s)**. Similarly, leave **Values (Y) in** set to rows because the population figures, which are plotted on the vertical or *y*-axis, are contained in a row. Make sure that the option **Categories (X Labels) in First Row** is checked.

6. Press the OK button and Excel will adjust the chart to include the new data values.

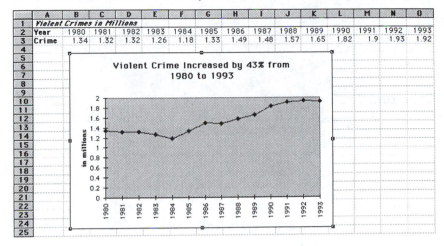

The title should also be changed so that the ending year is 1993 and the increase 43% (because $1.92/1.34 = 1.43$).

6.4 CREATING A CHART IN A CHART SHEET

So far all of the charts we have created have been embedded in a worksheet. Sometimes it is more convenient to produce a chart on a separate chart sheet. In this section you will learn how this is done. Open the worksheet **crime and population** from **datach6.xls**.

	A	B	C	D	E	F	G	H	I	J	K	L	M	N	O
1	*U.S. Population and Violent Crime*														
2	Year	1980	1981	1982	1983	1984	1985	1986	1987	1988	1989	1990	1991	1992	1993
3	U.S. Population	225	229	231	234	236	239	241	243	246	248	249	252	255	258
4	Violent Crime	1.34	1.32	1.32	1.26	1.18	1.33	1.49	1.48	1.57	1.65	1.82	1.9	1.93	1.92

Using Chart Wizard

Use Chart Wizard to plot the population data and place the chart in a separate chart sheet.

1. Select the range B2:O3.
2. Pull down the Insert menu, select Chart and then As New Sheet.
3. Excel will launch Chart Wizard. Make choices at the appropriate steps to produce the chart shown next. It will be contained in a new sheet.

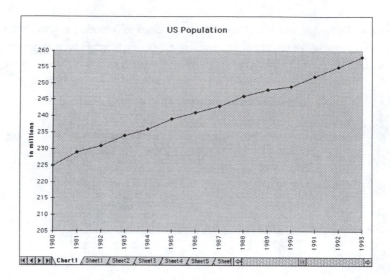

The name of the new chart sheet is **Chart 1**. You can see it in the line of tabs at the bottom of the window. (A chart sheet can be renamed or deleted just like a worksheet.) You can move from the chart sheet to any other sheet in the workbook by clicking on the appropriate tab.

Copying an Embedded Chart to a Chart Sheet

Sometimes you will want to copy a chart that you created and embedded in a worksheet into a separate chart sheet. You might do this if, for example, you wanted to use the chart to make a slide or overhead transparency.

1. Activate the embedded chart.
2. Click in a corner of the activated chart inside the border. Eight black rectangular handles will appear inside the border of the activated chart.
3. Pull down the Edit menu and release on Copy.
4. Press the F11 key, which you will find in the top row of your keyboard.
5. Pull down the Edit menu and release on Paste.

EXERCISES

The data for these problems is in **datach6.xls**.

Exercise 6.1 Produce three violent-crime charts discussed in the chapter: one that presents the data in a neutral way, one that exaggerates the increase, and one that makes the increase look quite small. Include the 1993 data. Arrange these charts embedded on a single worksheet and print it.

Exercise 6.2 Produce three different charts of the population data: one that presents it in a neutral way, one that exaggerates the increase, and one that makes the increase look quite small. Arrange these charts embedded on a single worksheet and print it.

Exercise 6.3 Open **workforce** from **datach6.xls**.
(a) Produce a vertical double bar chart of the data with the years on the horizontal axis and the percentages on the vertical axis. Title the chart.
(b) In the same worksheet embed a chart of the data in which the vertical axis begins at 30% and ends at 90%.
(c) In a text box compare these two charts. Do they convey different messages? Which is the most useful?
(d) Select the charts and text box and print them.

Exercise 6.4 Open the worksheet **500 m**.
(a) Produce a line chart that compares the times of men and women skaters. Use the default line; in other words, make no adjustment to the scaling.
(b) Produce a second chart and scale the vertical axis as shown next.

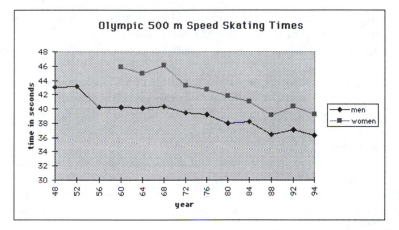

(c) In a text box compare these two charts. Do they convey different messages? Discuss the differences.
(d) Print the charts and text box.

Exercise 6.5 Use Chart Wizard to produce the following XY (Scatter) plot of violent crime versus population (crime on the vertical axis.) Begin by selecting the crime and population data, B3:O4, from the **crime and population** worksheet. Omit the years from your selection.

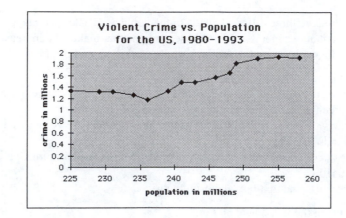

Open a text box and write a short paragraph commenting on the plot. What does the chart reveal about the relationship between crime and population? Print the scatter plot and text box.

Exercise 6.6 The number of violent crimes in 1993 was 43% greater than the number of violent crimes in 1982, as illustrated by the last chart in Section 6.3. This figure actually exaggerates the real increase in crime in American society. In evaluating the effect of violent crime on society, a much more useful and valid measure is one that compares the annual number of violent crimes with the population. Think about it this way: If the number of violent crimes is 1000 when the population is 2000, the crime rate is much more serious than if the population is a much larger number, say, 10,000. In the first instance the crime rate is 0.5 crimes per person and in the second it is the much smaller fraction 0.1 crimes per person.

(a) Calculate a new row in your **crime and population** worksheet that gives the number of violent crimes per person for each year.

(b) Since the numbers obtained in part (a) above are very small, create yet another row that gives the number of crimes per 100,000 people: Multiply the results from part (a) by 100,000.

(c) Expressed as a percentage, what is the change in crimes per 100,000 between 1980 and 1993?

(d) Create two charts from the data in part (b): one that represents the changes accurately and fairly and one that misrepresents the data in some fashion. Place the first chart in a chart sheet. Embed the second chart in the worksheet along with a text box in which you explain the misrepresentations. Why might some group or person be tempted to use this "dishonest" chart?

(e) Print the worksheet and the chart sheet.

WHAT TO HAND IN: Print no more than one page for each exercise assigned. Show plots and text boxes.

Investigating Associations

7

In This Chapter...

- XY Scatter Plots
- The coefficient of correlation
- Inserting a trendline
- The regression line
- Making estimates using the regression line
- Interchanging the axes
- Some nonlinear associations
- The effect of outliers on measures of association

Data is often investigated in pairs: height and weight, cigarette consumption and the incidence of lung cancer, the Consumer Price Index and interest rates, socioeconomic level and years of schooling, and so forth. The list could go on and on. In this chapter you will work with some of the tools available in Excel for the study of *bivariate* or *paired* numerical data.

7.1 CHARTING AN XY SCATTER PLOT

In this example we consider the association between homework averages and midterm examination scores. The scores of ten randomly selected students of statistics are shown next together with a scatter plot of the data.

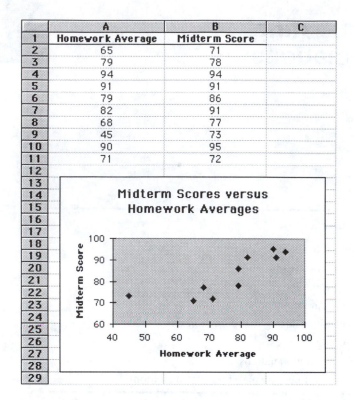

	A	B	C
1	Homework Average	Midterm Score	
2	65	71	
3	79	78	
4	94	94	
5	91	91	
6	79	86	
7	82	91	
8	68	77	
9	45	73	
10	90	95	
11	71	72	

In the scatter plot the two columns of data were combined in a special way to produce the scatter of points. Keep in mind that each point shown in the plot has a horizontal or x-coordinate and a vertical or y-coordinate. For example, the point on the far left of the scatter is (45,73). It has an x or horizontal coordinate of 45 and a y or vertical coordinate of 73. This point corresponds to the data in row 9 of the worksheet, which has the number 45 in the A column and 73 in the B column. The coordinates of each point in the scatter were produced in this same way by pairing a column A value with the column B value next to it. In this fashion, the ten data pairs in the worksheet produce the ten points in the scatter.

Using Chart Wizard to Draw a Scatter Plot in Excel 5 or 7

This section details the steps that produce the plot shown above. It will provide a template you can follow to generate a scatter plot for any set of bivariate data. If you are already an experienced Chart Wizard "pro," you probably can produce scatter plots on your own. In this case, skip to the next section. On the other hand, if you would like some help, begin by opening the worksheet **scores** in **datach7.xls**.

1. Select the two columns of data by clicking in cell A2 and dragging to cell B11.

2. Click the Chart Wizard button.

3. Click in your worksheet and drag to outline the region of the worksheet where you want the scatter plot drawn.

4. In Step 1 of Chart Wizard, make sure that the correct range for the bivariate data is entered.

5. In Step 2 of Chart Wizard, select the **XY (Scatter)** option.

6. In Step 3 of Chart Wizard, select option number 1.

7. In Step 4 of Chart Wizard, press the Next button.

8. In Step 5 of Chart Wizard, type in the Chart Title "Midterm Scores versus Homework Averages," and the Axis Titles "Homework Average" for **Category(X)** and "Midterm Score" for **Value(Y)**. Answer "No" to the **Add a legend?** question.

9. Click on the Finish button. The result will look something like the next picture.

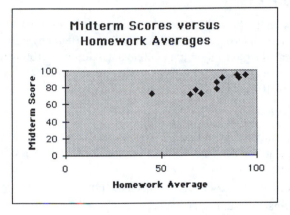

Notice that the data is clustered in the top right of the plot. In section entitled "Trimming the Axes" you will learn how to zoom in on this portion of the plot.

Using Chart Wizard to Draw a Scatter Plot in Excel 97

This section outlines the steps by which the scatter plot shown above can be produced in Excel 97. (For more detailed examples of Chart Wizard in Excel 97 see Appendix C.) Begin by opening **datach7.xls** and selecting the worksheet **scores**.

1. Select the two columns of data by clicking in cell A2 and dragging to cell B11.

2. Click the Chart Wizard button.

3. In Step 1 of Chart Wizard, click the **Standard Types** tab to bring these options forward and then the **XY (Scatter)** Chart Type. Click the top Chart sub-type—the one with no lines. Click the Next button.

4. In Step 2 of Chart Wizard, make sure that the data range is correct and that the **Series in Columns** option is selected. Click the Next button.

5. In Step 3 of Chart Wizard click on the **Titles** tab. Type in the Chart Title "Midterm Scores versus Homework Averages," and the Axis Titles "Homework Average" for **Category(X)** and "Midterm Score" for **Value(Y)**. Click on the **Legend** tab. If a check-mark appears in the **Show legend** box, click on it to remove the check. When the legend box is blank, click on the Next button.

6. In Step 4 of Chart Wizard, select the **As object in** option. Use the down arrow to place the worksheet name **scores** in this option box.

7. Click on the Finish button. The scatter plot will be drawn in the **scores** worksheet. The result will look something like the previous illustration.

Notice that the data is clustered in the top right of the plot. In the next section, "Trimming the Axes," you will learn how to zoom in on this portion of the plot.

Trimming the Axes

Most of the points on the plot are clustered at the upper right. The next steps will rescale the vertical and horizontal axes to zoom in on this cluster. Begin by trimming the horizontal axis so that it is scaled from a minimum of 40 to a maximum of 100 with major unit of 10. Then rescale the vertical axis so that it runs from a minimum of 60 to a maximum of 100 with a major unit of 10.

The Horizontal Axis

1. Double-click (single-click in Excel 97) inside the scatter plot chart to activate it. Eight black rectangular handles (and, in some versions of Excel a wide border) will appear around the chart.

2. Double-click on the horizontal axis to open the Format Axis dialog box.

3. When the Format Axis dialog box appears, click on the **Scale** tab and adjust the entries as shown next.

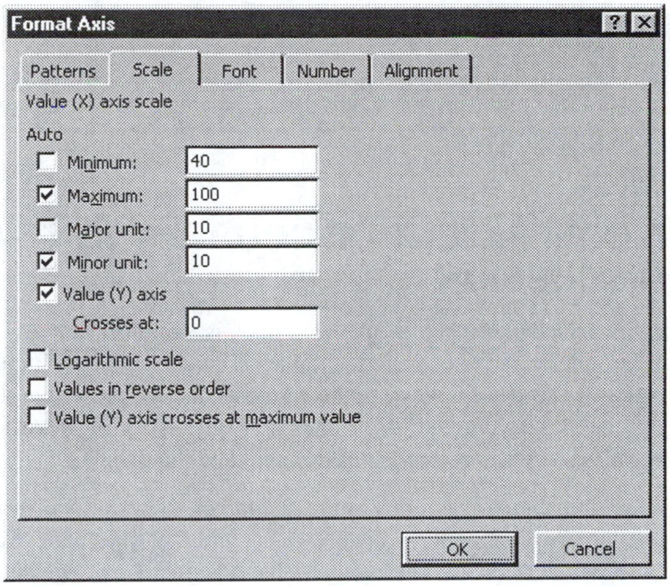

Format Axis [?] [X]

Patterns | Scale | Font | Number | Alignment

Value (X) axis scale

Auto

☐ Mi**n**imum: 40

☑ Ma**x**imum: 100

☐ Ma**j**or unit: 10

☑ Mi**n**or unit: 10

☑ Value (Y) axis

    Crosses at: 0

☐ **L**ogarithmic scale

☐ Values in **r**everse order

☐ Value (Y) axis crosses at **m**aximum value

[OK] [Cancel]

4. Click the OK button.

The Vertical Axis

1. Make sure the chart is activated and double-click on the vertical axis to again open the Format Axis dialog box.

2. Click on the **Scale** tab and adjust the entries as shown next.

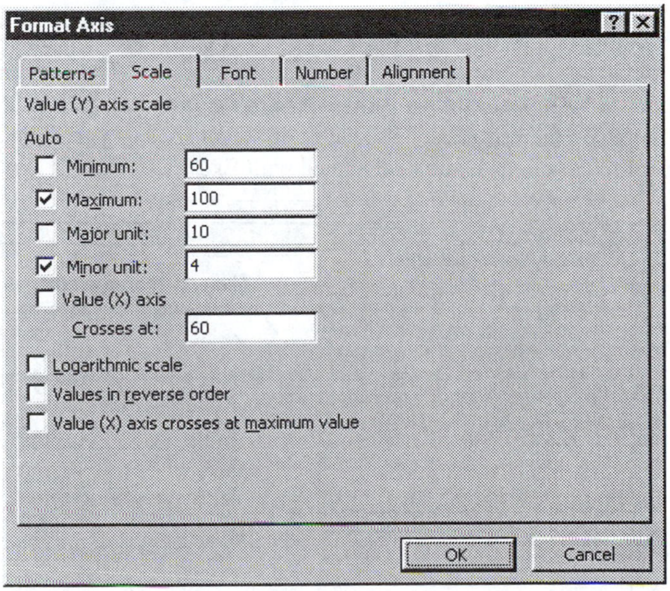

Format Axis [?] [X]

Patterns | Scale | Font | Number | Alignment

Value (Y) axis scale

Auto

☐ Mi**n**imum: 60

☑ Ma**x**imum: 100

☐ Ma**j**or unit: 10

☑ Mi**n**or unit: 4

☐ Value (X) axis

    Crosses at: 60

☐ **L**ogarithmic scale

☐ Values in **r**everse order

☐ Value (X) axis crosses at **m**aximum value

[OK] [Cancel]

3. Click on the OK button.

Your scatter plot should now look very much like the one shown at the beginning of this chapter. Be sure to save this scatter plot and the supporting data because they will both be used again later in this chapter.

7.2 THE COEFFICIENT OF CORRELATION—A VERY BRIEF DISCUSSION

The *correlation coefficient*, usually symbolized by r, is a number between -1 and 1 that measures the association between paired variables. The closer the coefficient is to 1 or to -1, the stronger the association between the variables and the closer the scatter of points is to a straight line. A zero correlation occurs when there is no linear association, either positive or negative, between the variables measured by the data.

A positive value for the coefficient means that the larger values of one variable tend to be paired with the larger values of the second variable and that the smaller values of each variable are also paired. This was the case with the homework average and midterm score data plotted in the previous section. A negative coefficient means that larger values of one variable tend to be paired with the smaller values of the other and vice versa.

Calculating the Coefficient of Correlation

The Excel command that calculates the coefficient of correlation has the form

$$=CORREL(x\text{-}range,y\text{-}range)$$

Study the example below, in which the coefficient of correlation is calculated for the association between midterm scores and homework averages. The scatter plot (notice that it has been repositioned) drawn for this data in the previous section shows a positive association between the variables: The scatter of points rises as you look at it from left to right. The coefficient of correlation calculated with Excel's CORREL command is 0.84.

In the exercises you will have the opportunity to study the scatter plots and coefficients of correlation for several data sets.

7.3 REGRESSION LINES OR TRENDLINES

When the coefficient of correlation is near 1 or −1, a line can be drawn through the data that, as you look at it from left to right, conforms fairly closely to the trend of the data points themselves. Such a line is displayed below. It is called a *trendline* in Excel.

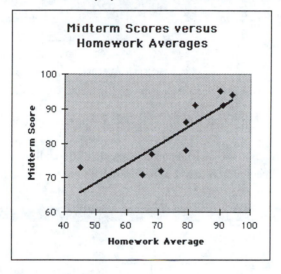

Interpreting a Trendline

In your statistics textbook a trendline is more likely to be referred to as a *regression line*. In either case it is the line that "best fits" the data points in the sense that the points lie closer to this line than they would to any other line drawn through the scatter of data points.

A regression line, when it fits the data well, is sometimes used to make predictions or estimates. For example, the line shown above seems to go through the point (85, 88). So, using the regression line as a predictor or model of the association between midterm performance and homework performance, we estimate that students with a homework average of 85 would score, on average, about 88 on the midterm examination. This estimate is rough because it was obtained by "eye-balling" the line. In the next sections, you will learn how to calculate such an estimate from the equation of the regression line.

How to Draw a Regression Line

Return to the scatter plot that you produced in the first section and activate the chart that contains it. Click on one of the data points. The points will change shape and color so that the scatter plot will look something like the one shown next.

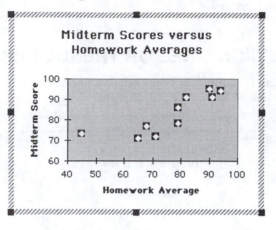

1. If you are working in Excel 5 or 7, pull down the Insert menu and release on Trendline... .

    In Excel 97 pull down the Chart menu and select Add Trendline... .

2. The Add Trendline dialog box will appear. Click on the tab labeled **Type** and then on **Linear** as shown next.

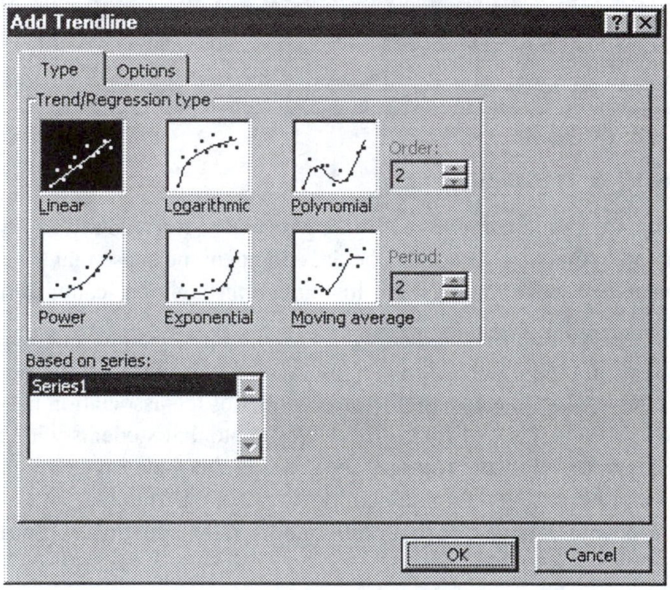

3. Click on the tab labeled **Options.** Make sure that **Display equation on chart** and **Display R-squared value on chart** are checked.

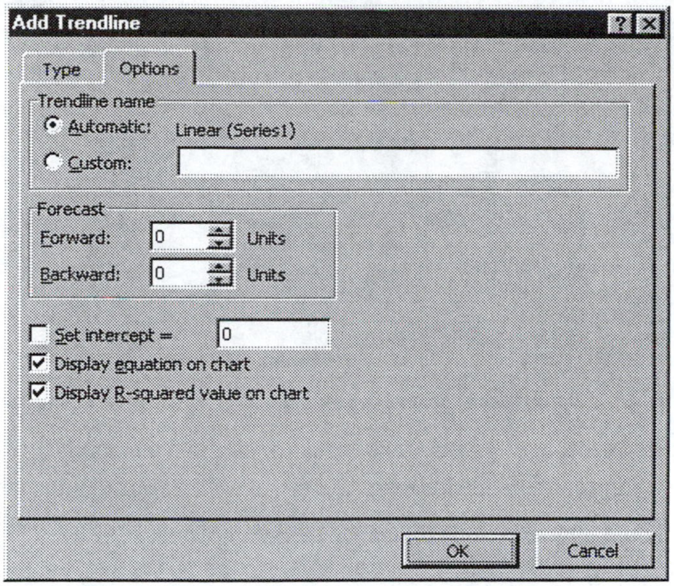

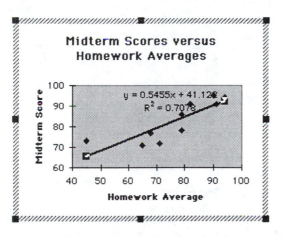

4. Click the OK button. Excel will draw the regression or trendline on the scatter plot as shown next.

Moving the Equations

If you like, you can neaten up the chart by moving the two equations off to the left a little so they don't obscure the scatter plot.

1. Activate the chart—make sure that the eight black handles are visible.
2. Click on the equations. They will be enclosed in a box.

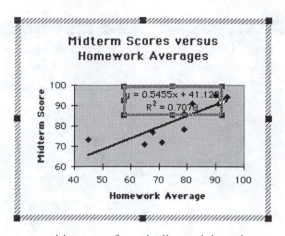

3. Drag the box to a new position away from the line and the points.

4. Deactivate the chart by clicking twice in the worksheet outside the chart boundary.

Your chart should now look something like the one shown next.

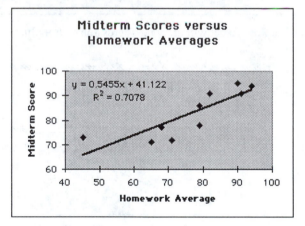

What the Equations Mean

The top equation, $y = 0.5455x + 41.122$, is the equation for the regression line. The second one, $R^2 = 0.7078$, gives the square of the coefficient of correlation.

$$R^2 = (0.8413)^2 = 0.7078$$

Refer to your statistics textbook for an explanation of the importance of this value. Stated very briefly, R^2 gives the proportion of the variation in the students' midterm scores that can be explained using the regression line model. So, in this example, approximately 71% of the differences in midterm scores from one student to the next can be explained by the differences in homework averages.

7.4 USING THE REGRESSION LINE TO MAKE ESTIMATES

Now that you have the equation

$$y = 0.5455x + 41.122$$

for the regression line, it can be used to estimate values for the y-variable (Midterm Score) when we are given a particular value for x (Homework Average). For example, a homework average of 85 corresponds to an estimated midterm score of approximately 87.87:

$$87.87 = 0.55(85) + 41.12$$

So, the "eye-ball" guess of 88 made earlier in the chapter was very close.

EXERCISES

The data for these exercises are in the workbook **datach7.xls**.

Exercise 7.1 Produce the Midterm Scores versus Homework Averages scatter plot together with the equations for the regression line and R^2 following the instructions given in Section 7.3. The data is in the worksheet entitled **scores**. Print the plot.

Exercise 7.2 Click on the tab **couples-age**.
(a) Produce a scatter plot of the data with husband's age on the horizontal axis and wife's on the vertical. Label the axes and the plot.
(b) Calculate the coefficient of correlation.
(c) Insert a trendline along with the equation for the regression line and for R^2.
(d) What wife's age does the regression line predict for a husband 30 years of age?
(e) Open a text box and write a paragraph discussing the direction and the strength of the association between these two variables.
(f) Print the calculations, scatter plot, and text boxes all organized on a single worksheet. Do not print the data.

Exercise 7.3 Click on the tab **couples-height**.
(a) Produce a scatter plot with trendline and R^2 equations. Make sure that the chart and its axes are labeled.
(b) Calculate the coefficient of correlation.
(c) What wife's height does the regression line predict for a husband 1700 millimeters tall?
(d) Open a text box and write a paragraph discussing the direction and the strength of the association between these two variables. Compare the association of the height

variable with the association found in Exercise 7.2 for the age variable.

Exercise 7.4 Click on the tab **Ht-GPA**.

(a) Produce a scatter plot with trendline and R^2 equations. Make sure that the chart and its axes are labeled. Trim the axis if needed.

(b) Calculate the coefficient of correlation.

(c) Open a text box and write a paragraph discussing the direction and the strength of the association between these two variables.

(d) Print the calculations, scatter plot, and text boxes all organized on a single worksheet. Do not print the data.

Exercise 7.5 Click on the tab **SAT-GPA**.

(a) Produce a scatter plot with trendline and R^2 equations. Make sure that the chart and its axes are labeled. Trim the axes if needed.

(b) Calculate the coefficient of correlation.

(c) Open a text box and write a paragraph discussing the direction and the strength of the association between these two variables.

(d) Print the calculations, scatter plot, and text boxes all organized on a single worksheet. Do not print the data.

Exercise 7.6 Click on the tab **sleep&ges**.

(a) Produce a scatter plot with trendline and R^2 equations. Make sure that the chart and its axes are labeled. Trim the axes if needed.

(b) Calculate the coefficient of correlation.

(c) Open a text box and write a paragraph discussing the direction and the strength of the association between these two variables.

(d) Print the calculations, scatter plot, and text boxes all organized on a single worksheet. Do not print the data.

Exercise 7.7 Open the worksheet **health**. Produce a scatter plot of male and female lifespans with trendline and R^2 equations. Edit the axis to eliminate any empty space. Calculate the coefficient of correlation. Open a text box and write a paragraph discussing the direction and the strength of the association between these two variables. Print the calculations, scatter plot, and text boxes all organized on a single worksheet. Do not print the data.

Interchanging the Axes

Exercise 7.8 Open the worksheet **health**. In this exercise a scatter (X-Y) plot of lifespan versus number of people per doctor will be drawn (in other words, a scatter in which lifespan is on the vertical axis and people per doctor on the horizontal). This is a bit awkward because the lifespan data is in column B of the worksheet and the people per doctor data is in column D, and Chart Wizard will automatically use data in the earlier column (B comes before D) for the horizontal axis. Follow the steps below to reverse this default order.

(a) Select the data and draw the default scatter plot.

(b) Activate the chart and select the points. They will change color and shape as shown next.

S1	▼	=SERIES(,health!$B$2:$B$39,health!$D$2:$D$39,1)			
	A	**B**	**C**	**D**	**E**	**F**

	A	B	C	D	E	F
1	Country	Life Span	People per television	People per Doctor	Female life span	Male life Span
2	Argentina	70.5	4	370	74	67
3	Bangladesh	53.5	315	6166	53	54
24	Peru					62
25	Philippines					62
26	Poland					69
27	Romania					69
28	Russia					64
29	South Africa					61
30	Spain					75
31	Sudan					52
32	Taiwan					72
33	Thailand					66
34	Turkey					68
35	Ukraine					66
36	United Kingd(					73
37	United States					72
38	Venezuela					71
39	Vietnam	65	29	3096	67	63

(c) The formula bar contains a SERIES command. Edit this command changing all of the B's to D's and the D's to B's so that it reads as follows.

$$=\text{SERIES}(,\text{health!}\$D\$2{:}\$D\$39,\text{health!}\$B\$2{:}\$B\$39,1)$$

(d) Press the ENTER key. The axes will be interchanged so that the chart now looks like the illustration shown next.

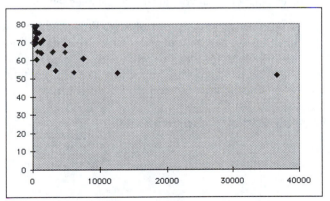

(e) Insert the trendline and R^2 equations. Edit the axis to eliminate any empty space. Calculate the coefficient of correlation.

(f) Open a text box and write a paragraph discussing the direction and the strength of the association between these two variables.

The Importance of the Scatterplot

Exercise 7.9 Open the worksheet labeled **Anscombe Data**. For each of the following ranges, draw a scatter plot and insert its trendline, the equation of the trendline, and the value of R^2.
(a) B3:C13
(b) B3:B13 and D3:D13
(c) B3:B13 and E3:E13
(d) F3:G13
(e) In which of the plots, if any, does the trendline fit the scatter?
(f) If you were given only the equations of the regression lines and the values of R^2 for these data sets and had no access to the scatter plots, what erroneous conclusions might you draw?
(g) Print the four plots and your answers to (e) and (f) all organized nicely on a single page.

Not All Data Is Linear

Exercise 7.10 Sometimes it makes no sense to use a linear equation to model data. For a simple example, click on the tab **non-linear**.
(a) Produce a scatter plot, with column A providing the x-axis data and column B the y, with linear trendline and R^2 equations.
(b) Calculate the coefficient of correlation.
(c) Make a second scatter plot of the data and this time insert a *polynomial* trendline.
(d) What is the coefficient of correlation for the polynomial plot?
(e) Print the plots, calculations, and your answer to the question.

Exercise 7.11 Open the worksheet called **investment**. Produce three scatter plots, with column A providing the x-axis data and column B the y. Insert three different trendlines—linear, polynomial, and exponential—in the scatter plots. Open a text box and explain which of the three types of trendlines best fits the data and why you think so. Print the plots and discussion on a single sheet of paper.

Outliers Play Havoc with Correlation

In the next two exercises you will have the opportunity to look at two sets of data in which outlying data points dramatically affect the strength of the association.

Exercise 7.12 Click on the tab labeled **smoking**.

Create a spreadsheet like the one shown next.

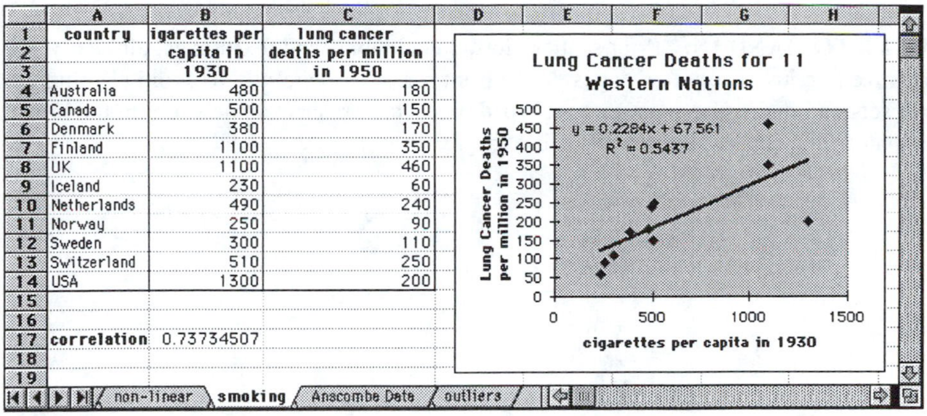

	A	B	C	D	E	F	G	H
1	country	igarettes per	lung cancer					
2		capita in	deaths per million					
3		1930	in 1950					
4	Australia	480	180					
5	Canada	500	150					
6	Denmark	380	170					
7	Finland	1100	350					
8	UK	1100	460					
9	Iceland	230	60					
10	Netherlands	490	240					
11	Norway	250	90					
12	Sweden	300	110					
13	Switzerland	510	250					
14	USA	1300	200					
15								
16								
17	correlation	0.73734507						
18								
19								

Chart: **Lung Cancer Deaths for 11 Western Nations**, with $y = 0.2284x + 67.561$, $R^2 = 0.5437$, y-axis "Lung Cancer Deaths per million in 1950", x-axis "cigarettes per capita in 1930".

Tabs: non-linear \ smoking / Anscombe Data / outliers

Study the effect of the USA data point on the association between the variables.

(a) Select the range B14:C14. Select Clear and then Contents from the Edit menu. Notice the changes in the trendline and the value of R^2. Select Undo from the Edit menu to return the USA data point. Repeat the removal and reentry of the USA data point until you understand its effect on the measures of association.

(b) Open a text box, write a paragraph describing the effect that removing the USA point from the data set has on the measures of association between the two variables, and print the text box.

Exercise 7.13 Open the worksheet containing the life span and gestation period data for several dozen animal species. It is called **lifesp&ges**.

(a) Set up a worksheet for this data like the one illustrated in the last exercise.

(b) Identify two or three animals that seem to have outlier status. Study the effect of each of the animals you identified on the measures of association and write up your findings in a text box. Print the text box.

Exercise 7.14 Open the worksheet entitled **O-ring failures**. Read the text box describing the data. Produce a scatter plot of O-ring failures vs. temperature with trendline and R^2 equations.

(a) using all of the data.

(b) using only the data which includes O-ring failures, rows 18 through 24.

(c) Open a text box and write a paragraph discussing and comparing the direction and the strength of the association between these two variables in parts (a) and (b).

(d) At 11:38 am on January 28, 1986, when the ill-fated space shuttle Challenger was launched at the Kennedy Space Center, the temperature was 32 degrees Fahrenheit. Based on the calculations you made in parts (a) and (b) above, would you have voted to abort the launch? Why or why not.

(e) Print the calculations, scatter plot, and text boxes all organized on a single worksheet. Do not print the data.

WHAT TO HAND IN: Print scatter plots, calculations, and text boxes, all neatly organized on just a very few pages. Make print selections carefully to avoid printing the data sets and make sure your name is typed on each worksheet somewhere in the print region (or in a header, if you wish to use this feature).

Investigating the Significance of *r*

In This Chapter...

- RAND
- Calculation modes
- Fill Down and Fill Series
- Plotting a random scatter
- Randomizing a list
- *r* and sample size
- More on outliers
- The Infamous 1970 Draft Lottery

The word *significant* has a special meaning in statistics: it is used to describe an event not easily explained by chance factors alone. For example, no one thinks twice about a coin that comes up heads on a single toss. After all, heads is expected about half the time. But fifteen heads in a row in fifteen tosses is another matter entirely. It is so unlikely (the probability is less than 1 in 30,000) that if you witnessed such an event, you would suspect a trick and insist on inspecting the coin to see if it had two heads or was somehow weighted. It is not that fifteen heads in fifteen tosses never occurs by chance. It could, but the event is rare, so when it happens it is reasonable to at least consider causes other than the random behavior of a fair coin. An occurrence, such as the long run of heads just discussed, too improbable to be satisfactorily explained by chance alone is referred to as a *statistically significant* event.

In the last chapter you studied paired variables using the coefficient of correlation or *r*-value to measure the strength of the association between the variables. This chapter introduces the question of the statistical significance of *r*. Suppose, for example, that the coefficient of correlation for 10 pairs of data values turns out to be $r = 0.5$. The value suggests a modest association between the variables to be sure, but is the value *significant*? Could an *r*-value as high or higher than 0.5 be explained by chance alone? To begin to answer these questions we need some hands-on experience with random scatters, with the behavior of numbers paired together completely by chance. The Excel features introduced in the next section provide just the tools needed.

8.1 SOME USEFUL EXCEL FEATURES

RAND

The Excel function RAND selects a number between 0 and 1 at random. In the next illustration, Excel returned the number 0.454516 when =RAND() was entered. Notice that this command has no argument. The parentheses must be typed in, but they are left empty.

A1		=	=RAND()
A	**B**	**C**	**D**
1	0.454516		
2			

Open a new workbook and enter the command =RAND() into cell A1. You will almost certainly get a different value than the one shown above. Why?

The command RAND can be combined with others in order to change the interval from which the random number is selected. For example, in the next illustration RAND is used to generate a random number between 3 and 6.

A1		=	=3*(1 + RAND())
A	**B**	**C**	**D**
1	3.033544		
2			

How to Recalculate all the Functions in a Worksheet: SHIFT-F9

With the previous worksheet still active, hold down the SHIFT key and press the F9 key several times. Notice that Excel recalculates the RAND command every time SHIFT-F9 is pressed. In general, SHIFT-F9 causes Excel to recalculate every function in the active worksheet. This feature will be of great value in studying random phenomena. (NOTE: If you press only the F9 key without holding down the SHIFT key, Excel will recalculate not only the active worksheet, but every other worksheet in the open workbook as well. If the workbook contains many sheets, this can be a very time consuming process. Use SHIFT-F9 for the purposes of recalculation unless you have a *very* good reason for doing otherwise.)

How to Change the Recalculation Mode

When you enter a function into a worksheet and copy it to a new cell, as with a black cross drag, Excel automatically recalculates the function. This recalculate feature of Excel can be turned off, and when using Excel to generate random numbers, you will want to do this. The reason is that RAND has the peculiar feature that it is recalculated whenever a command is entered anywhere in the active workbook. The result is that a workbook containing many RAND commands is frequently recalculated, a process that greatly slows down all of the rest of the operations. As it turns out, very little is lost when the recalculate feature is disabled, since SHIFT-F9 can always be used to force the calculation of worksheet formulas.

Before going on to the next section, follow the procedure below to change the Calculate feature of Excel from Automatic to Manual.

1. Pull down the Tools menu and select Options... .
2. In the Options dialog box, click on the tab labeled **Calculation**.
3. Under **Calculation** select **Manual**. The result will look like the next picture.

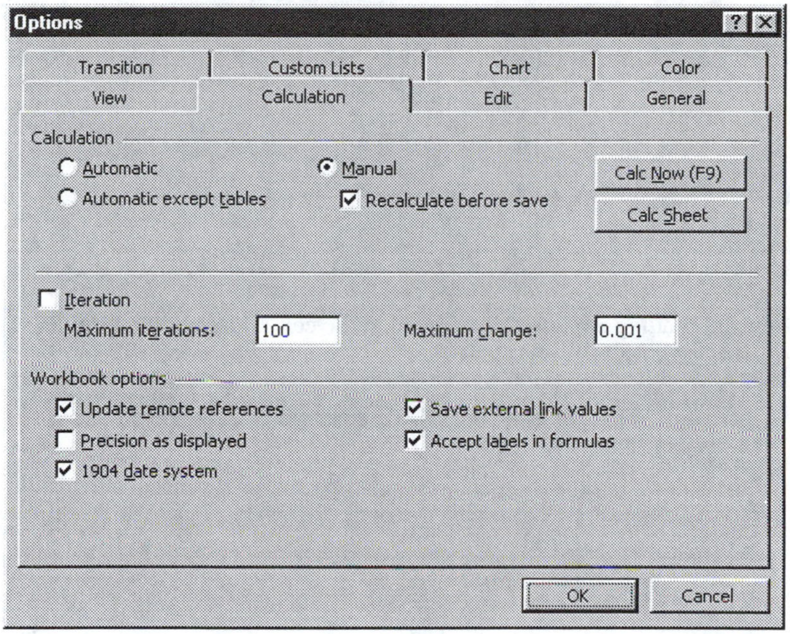

4. Click the OK button.

Whenever you wish you can return to the Automatic calculation mode, open this same dialog box and select **Automatic** under **Calculation**.

8.2 HOW TO DRAW A RANDOM SCATTER OF POINTS

This section outlines a procedure for generating a random scatter of points. In the example you will draw a scatter plot and trend line for ten randomly generated points. The procedure can be easily adjusted to produce any number of such points.

1. Open a fresh worksheet.
2. Make sure that the **Calculation** option is set to **Manual**.
3. Type the labels "x" and "y" into cells A1 and B1.
4. Enter the command =RAND() into cell A2.
5. Copy the command to cell B2 using a black cross drag.
6. Select cell A2, hold down the SHIFT key and click in cell B11.
7. Pull down the Edit menu and select Fill Down.
8. Hold down the SHIFT key and press F9. You should now have ten pairs of randomly selected numbers looking something like the next illustration.

	B11		= =RAND()	
	A	B	C	D
1	x	y		
2	0.953018	0.356788		
3	0.580398	0.227318		
4	0.304084	0.6989		
5	0.624541	0.104151		
6	0.825903	0.073947		
7	0.407794	0.44383		
8	0.155898	0.063389		
9	0.288597	0.754572		
10	0.519843	0.860978		
11	0.592989	0.87785		
12				

9. Enter the heading **r =** into cell A13 and in the cell B13 enter the formula =CORREL(A2:A11,B2:B11).

	B13		= =CORREL(A2:A11,B2:B11)		
	A	B	C	D	E
1	x	y			
2	0.953018	0.356788			
3	0.580398	0.227318			
4	0.304084	0.6989			
5	0.624541	0.104151			
6	0.825903	0.073947			
7	0.407794	0.44383			
8	0.155898	0.063389			
9	0.288597	0.754572			
10	0.519843	0.860978			
11	0.592989	0.87785			
12					
13	r =	-0.21431			
14					

10. Use Chart Wizard to draw a scatter plot of the ten random points and insert a trendline. (Detailed instructions for completing this step are given in Section 7.1.)

Your worksheet should resemble the one shown next. Since the process is random, your points and r-value will, of course, be different.

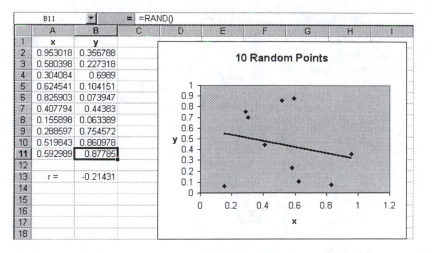

Save this workbook as you will need it for one of the exercises at the end of the chapter.

8.3 RANDOMIZING A LIST

In Exercises 8.5 and 8.6 you will be asked to investigate the fairness of the 1970 draft lottery. The investigation requires the random rearrangement or shuffling of the dates of the year. In this section we provide an example of an Excel shuffling technique that you can adapt to the lottery problem.

Suppose you have the task of assigning ten people to experimental treatments, and you must make sure that the order in which the ten are assigned is random. With such a small group, it would be easy to design a method not requiring Excel. For example, you could assign each person a number from 1 to 10 and then write the numbers on 10 slips of paper, place the slips in a paper bag, shake the bag up well, and draw out the numbers one at a time. If the number three were drawn first, then the individual with that number would get the first treatment assignment. Continuing until all ten slips were drawn would complete the treatment assignments. This method is fine in theory, but has two drawbacks in practice. The first is the physical problem of shaking the sack sufficiently to thoroughly shuffle all of the slips. This is not as simple as it sounds. The second difficulty is that this method becomes quite cumbersome when a large number of assignments must be made. In the Excel-based procedure the shuffling of the assignments is handled with the RAND command, which eliminates the physical problem. As you will see, it is a method that can be easily applied to the randomization or shuffling of much longer lists.

How to Shuffle the List 1,2,3,4,5,6,7,8,9,10

1. Begin by entering ten randomly selected numbers in the range A1:A10.
    (a) Set the **Calculation** mode to **Manual**.
    (b) Open a worksheet and enter the command =RAND() into cell A1.

(c) Click in cell A1, hold down the SHIFT key and click in cell A10.

(d) Select Fill Down from the Edit menu.

(e) Hold down the SHIFT key and press the F9 key.

The result should look something like the next picture.

	A10			=	=RAND()		
	A		B	C		D	
1	0.00273						
2	0.5272						
3	0.84659						
4	0.572019						
5	0.006007						
6	0.20413						
7	0.972754						
8	0.630413						
9	0.493606						
10	0.919641						
11							

2. List the numbers from 1 to 10 in the range B1:B10.

   (a) Enter the number 1 into cell B1.

   (b) Select the range B1:B10: click on cell B1, hold down the SHIFT key, and click on cell B10.

   (c) From the Edit menu select Fill Series... .

   (d) When the Series dialog box opens, click OK.

Your worksheet should now show the numbers from 1 to 10 in column B paired with the random list of ten numbers in column A.

	A	B
1	0.002730127	1
2	0.527200414	2
3	0.846590031	3
4	0.572018601	4
5	0.006006802	5
6	0.204130071	6
7	0.972754209	7
8	0.630412614	8
9	0.493606386	9
10	0.919640709	10

3. The next step is to "shuffle" the list 1,2,3,4,5,6,7,8,9,10. This is done by arranging the ten pairs in order of the random numbers.

   (a) Click in cell A1, hold down the SHIFT key and click in cell B10. Make sure that the cell A1 is the active cell as pictured next.

A1		= =RAND()	
A	**B**	**C**	

	A	B	C
1	0.002730127	1	
2	0.527200414	2	
3	0.846590031	3	
4	0.572018601	4	
5	0.006006802	5	
6	0.204130071	6	
7	0.972754209	7	
8	0.630412614	8	
9	0.493606386	9	
10	0.919640709	10	
11			

(b) Click the Sort Ascending button on the Toolbar. Your worksheet will look something like the one shown next.

	A	B
1	0.02390049	2
2	0.065244106	8
3	0.543921588	7
4	0.615345499	1
5	0.724319895	4
6	0.757015405	5
7	0.841503311	10
8	0.859620468	6
9	0.938042256	3
10	0.973692106	9

The numbers from 1 to 10 are now sorted randomly in column B. The listing tells us that person number 2 should be assigned first to the treatment, person number 8 second, person 7 third, and so forth. The random numbers in column A are no longer needed, so replace them with the numbers from 1 to 10 indicating the order of treatment.

(c) Click in cell A1, hold down the SHIFT key and click in cell A10. Pull down the Edit menu and select Fill Series... . Click the OK button in the Series dialog box.

(d) Shown next is an illustration of the final form of the assignment.

	A	B
1	1	2
2	2	8
3	3	7
4	4	1
5	5	4
6	6	5
7	7	10
8	8	6
9	9	3
10	10	9

EXERCISES

The exercises below will use material contained in **datach8.xls**.

Experimenting with Sample Size: How Does *n* Affect *r*?

Exercise 8.1 Create a random shuffle of the numbers from 1 to 10 and print it.

Exercise 8.2 There can be no *real* association between the variables in a random scatter. After all, the numbers are paired up using a random process. This means that the coefficient of correlation or r-value obtained for such a scatter results from chance factors alone. Experience with random scatters will help you to interpret the significance of the correlations you encounter for real data. Before you can conclude that a particular correlation indicates an actual association between variables, you must be reasonably sure that something other than chance is at work. In order to make this judgment it helps to have some experience with random scatters, in particular, with the effect of the sample size on the variation in r-values.

For each of the next four exercises, draw a random scatter of the indicated number of points following the instructions in Section 8.2. Draw the scatter plot, and insert a trendline. Calculate the coefficient of correlation for the points as described in the section. Use SHIFT-F9 to recalculate the worksheet 20 times, keeping track of the r-values you obtain.
  (a) Two points. What values do you get for r and why?
  (b) Three points. What is the range of r-values? What percent of the 20 r-values lie within 0.1 of zero, that is, between $-.1$ and $.1$? What percent lie further away from zero than 0.5?
  (c) Ten points. What is the range of r-values? What percent of the 20 r-values lie within 0.1 of zero? What percent lie further away from zero than 0.5?
  (d) 100 points. What is the range of r-values? What percent of the 20 r-values lie within 0.1 of zero? What percent lie further away from zero than 0.5?
  (e) Print your calculation and discussions.

Exercise 8.3 Enter the 20 r-values you obtained in Exercise 8.2 for 3, 10, and 100 points into the A, B, and C columns of a new worksheet.
  (a) Construct parallel boxplots for the data.
  (b) Open a text box and discuss the results of your calculations. How does sample size seem to effect the size and variability of r? For which of the sample sizes would an r-value of 0.6, for example, probably indicate a significant association?
  (c) Print the boxplot and text box.

Exercise 8.4 Open the worksheet called **outliers** in **datach8.xls**.
  (a) How many points are there in the random scatter, and what are the ranges for their x and y coordinates? How many points lie outside the scatter and what are their coordinates?

(b) Click in cell B105. What range of points are included in this correlation? Answer the same question for B108.
(c) Recalculate the worksheet 20 times. Compare the two correlations. Open a text box and describe the effect of the outliers on the correlation. Print the text box.

The 1970 Draft Lottery

A generation ago young American men were subject to the military draft. Prior to 1970 certain categories of men, such as those enrolled in college, could obtain deferments from their draft boards and so postpone military service and in many instances avoid it altogether. In 1970 these deferments were eliminated and a lottery instituted in an effort to make all young men equally vulnerable to the draft. The lottery was designed as follows: Each eligible young man was assigned a number between 1 and 366 that corresponded to his birth date. January 1 was birthday number one and subsequent days of the year (including February 29) were assigned numbers in order until December 31, which was birthday number 366. Each of the numbers from 1 to 366 was inserted into a capsule. The capsules were mixed in a rotating drum and then drawn one at a time. The first number drawn was 305, which meant that men with the corresponding birth date, October 31, were drafted first. The order of birth dates for the draft lottery is found in the worksheet **draft lottery** in the workbook **datach8.xls**. Open this worksheet.

Exercise 8.5 In a fair lottery there should be no predictable pattern in the order of the birth dates.
(a) Draw a scatter plot of the data in columns A and B with birth date plotted on the horizontal axis and lottery number on the vertical. Insert the trendline.
(b) Calculate the coefficient of correlation for the data. You should get -0.23 rounded to two decimal places. What does a negative association mean for this data?
(c) The designers of the lottery would have preferred a correlation nearer zero. Why?
(d) Was the ordering of birth dates really random? Answer the question this way: Create a truly random ordering of the numbers 1,2,3, ..., 365, 366 using the shuffling method described in Section 8.3. Find the correlation of these "lottery" numbers vs. the birth dates. The result will look something like the worksheet shown next, with the birth dates in column A and the lottery numbers in column B.

B368		=CORREL(A1:A366,B1:B366)	
	A	B	C	D
1	1	131		
2	2	290		
3	3	343		
4	4	120		
363	363	95		
364	364	149		
365	365	360		
366	366	350		
367				
368	correlation	0.019268644		
369				

Print a copy of your lottery simulation.

Exercise 8.6 The r-value shown in Exercise 8.5 is near zero, as one would expect for such a large number of randomly paired values. (See the first five exercises.) Now, open the worksheet entitled **dynamic lottery**. A picture of this worksheet is shown next.

	A	B	C	D	E
1	1	25			
2	2	290			
3	3	57			
4	4	341			
355	355	115			
356	356	176			
357	357	364			
358	358	145			
359	359	53			
360	360	43			
361	361	221			
362	362	7			
363	363	75			
364	364	205			
365	365	202			
366	366	186			
367					
368	correlation	-0.07053372			

draft lottery \ **dynamic lottery**

It contains a lottery simulation very much like the one you just did, but there is an important difference. The simulation has been automated so that you can make random lottery assignments over and over again observing the change in the correlation. The procedure works as follows.
1. Click in cell A1.
2. On the Macintosh hold down the OPTION key and the Open-Apple key and press the l (lowercase L) key. On a Windows machine hold down the CTRL key and press the lowercase L key.

Run the lottery simulation 30 or 40 (more if you like) times, keeping track of the correlations you obtain. What is the range of your correlations? What fraction of them differ from zero by as much as 0.23 or more? Draw a boxplot of the correlations. Do you think it is likely that the 1970 lottery was truly random? Why or why not? Print your calculations and discussion.

WHAT TO HAND IN: Print the worksheet selections asked for and make the selections carefully. Don't forget to type your name in the print region.

Normal Distributions

9

In This Chapter...

- Drawing bell-shaped curves
- Calculating normal probabilities
- NORMDIST
- NORMINV
- The Drawing Toolbar
- NORMSDIST
- NORMSINV
- Drawing normal approximations to histograms
- The Empirical Rule
- PERCENTILE

A normal curve is bell-shaped and symmetric and is described by its mean and standard deviation. In this chapter you will learn how to use Excel to draw normal curves and to calculate proportions and percentiles associated with normally distributed populations. Tools for deciding whether a set of data is normally distributed will also be introduced.

9.1 PLOTTING A NORMAL CURVE

In 1995 the College Board revised the scoring of the Scholastic Aptitude Test (SAT) so that it now has a mean of 500 and a standard deviation of 100. Scores on this exam have a bell-shaped distribution, a picture of which is displayed next.

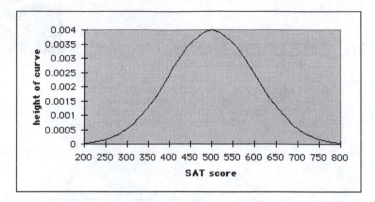

This curve is called a *density function* for the SAT scores. This is a technical term meaning that the curve was mathematically designed, using techniques from the calculus too advanced to go into here, so that the area between the curve and a range of scores on the horizontal axis equals the proportion of students who received those scores. For example, the total area under the curve is 1 which means that all of the SAT scores fell between 200 and 800. Half the area under the curve lies to the right of 500, so 50% of the students scored between 500 and 800 and the other 50% between 200 and 500.

Follow the instructions below to draw a normal curve having a mean of 500 and a standard deviation of 100.

Calculate a List of Points

1. Open a new workbook. Make sure that the Calculation mode is set at Automatic. (See Section 8.1 for instructions.)
2. Type the name of the variable, in this case "SAT score," in cell A1 and "height of curve" in cell B1. Adjust the column width as needed.
3. Split the screen horizontally so that rows 1 and 32 are both visible.
4. In column A generate a list of values for the variable, which begins 3 standard deviations below the mean and ends 3 standard deviations above the mean. (Recall that with a normally distributed variable virtually 100% of the data will lie within 3 standard deviations of the mean.) The increment from one value in the list to the next should be about one-fifth of the standard deviation in order to produce a nice smooth bell.
    (a) In cell A2 enter the number that lies 3 standard deviations below the mean, 200 in this example. Press the ENTER key.
    (b) Select cell A2, pull down the Edit menu, select Fill and then Series... .
    (c) When the Series dialog box appears, select the options as illustrated next.

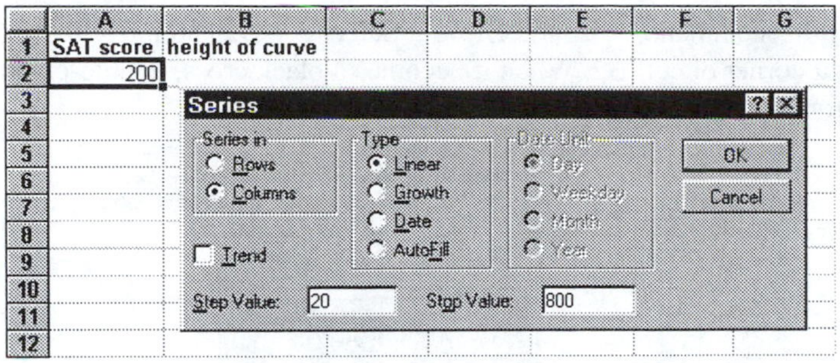

Make sure that **Series in...** has the **Columns** setting, that **Step Value** is set at **20** (one-fifth of the standard deviation), and that **Stop Value** is set at **800** (the mean plus 3 standard deviations).

(d) Click the OK button. A sequence of SAT values will appear in column A as shown next.

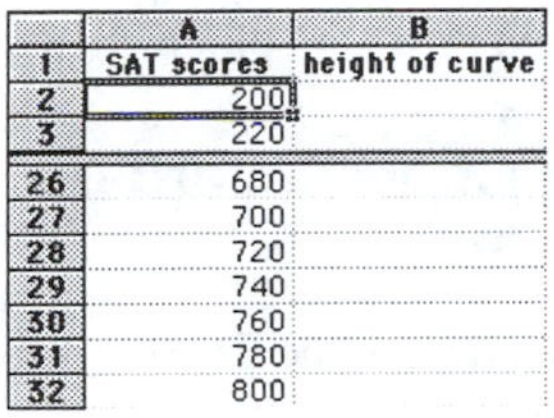

5. In cell B2 calculate the height of the normal curve above 200: Enter the command =NORMDIST(A2,500,100,0) into cell B2. The result is shown next.

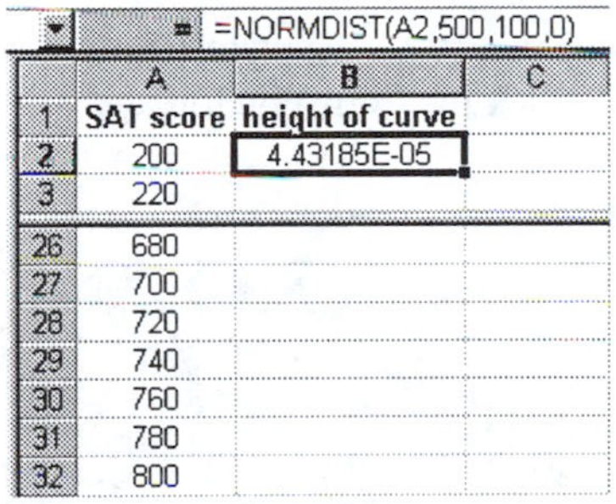

In general, the height of a normal curve with a given mean and standard deviation above the number x on the horizontal axis is given by the command =NORMDIST(x, mean,standard deviation,0).

Note: When the fourth argument in the command is set at 1, instead of 0, Excel will return the area under the normal curve to the left of x rather than the height of the curve at x.

6. Fill this command down to cell B32. Select cell B2 and move the cursor to the lower right corner of cell B2. When it becomes a black cross, double-click. The Fill will be automatically executed with the result shown next.

	= =NORMDIST(A2,500,100,0)	
	A	B	C
1	SAT score	height of curve	
2	200	4.43185E-05	
3	220	7.91545E-05	
26	680	0.000789502	
27	700	0.00053991	
28	720	0.000354746	
29	740	0.000223945	
30	760	0.00013583	
31	780	7.91545E-05	
32	800	4.43185E-05	
33			

7. The values paired in columns A and B are points on the SAT density function, its bell-shaped curve. The next step is to plot the curve.

Plot the Points Using Chart Wizard in Excel 5 or 7

1. Select the range A2:B32 and click the Chart Wizard button.
2. At Step 2 select **XY (Scatter)**.
3. At Step 3 select **Option 6**.
4. At Step 4 click the Next button.
5. At Step 5 select No Legend. In **Category(X)** type "SAT score" and in **Value(Y)** type "height of the curve." Click the Finish button and Excel will draw the normal curve as shown next.

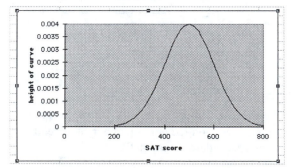

Plot the Points Using Chart Wizard in Excel 97

1. Select the range A2:B32 and click the Chart Wizard button.
2. At Step 1 select **XY(Scatter)** and the plain curved line option.
3. At Step 2 select Next.

4. At Step 3 remove the legend and the gridlines. Enter the titles. In **Label Category(X)** type "SAT score" and in **Values(Y)** type "height of the curve."

5. At Step 4 select the **As object in worksheet** option. Click the Finish button and Excel will draw the normal curve as shown in the previous picture.

Editing the Curve

At this point the horizontal axis will need some cleaning up. Double-click on the chart to activate it (single-click in Excel 97) and then click on the horizontal axis. Pull down the Format menu and select Selected Axis... . When the dialog box appears, click on the tab labeled **Scale**. Set **Minimum** at **200**, **Maximum** at **800**, and **Major Unit** at **50**. Click the OK button. The normal curve will now look something like the one displayed next.

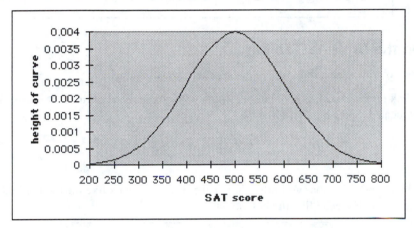

9.2 USING THE DRAWING TOOLBAR TO INSERT A VERTICAL LINE

It is often helpful to draw vertical lines between a normal curve and the horizontal axis because such lines can be used to set off the area under the curve corresponding to an interval of scores. For example, the vertical line drawn in the chart below at the score 580 divides the area under the curve into two regions. The area of the region to the left of the line gives the proportion of students who scored below 580 on the SAT exam. The region on the left of the line gives the proportion who scored above 580.

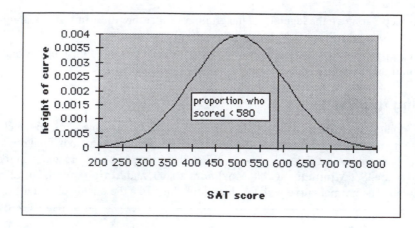

Open the Drawing Toolbar

In order to draw such a line, follow the next set of instructions.

1. Click on the Drawing Tool button, which will look like one of the next illustrations. It is located on the Standard Toolbar.

2. The Drawing Toolbar will open. In Excel 5 or 7 it will look like the top bar in the next picture, in Excel 97 like the bottom bar.

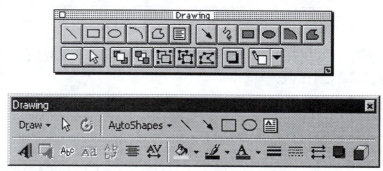

Draw the Line

1. Click the Line Button which is illustrated next. The cursor will become a crosshair.

2.   Center the crosshair on the horizontal axis at 580 and press and hold the mouse button while dragging straight up until the crosshair is centered on the curve.

3.   Release the mouse button and the vertical line will appear with small black boxes at either end as shown next.

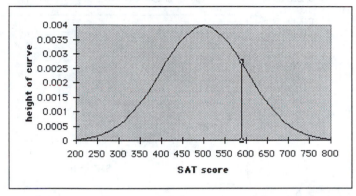

If you don't like the looks of the line, press the DELETE key and start over. If the line seems OK, click outside the chart to remove the black boxes. If you now wish to delete the line, click on it to select it (the black boxes will reappear) and press the DELETE key.

A text box that identifies the region to the left of the vertical line can be inserted in the usual way using the Text Box button. (In Excel 97 there are other methods available for attaching text to a chart, which you might like to explore on your own. For example, try clicking the slanting A button on the Drawing Toolbar. Meanwhile, text boxes have the advantage of familiarity.)

9.3 CALCULATING PROPORTIONS AND PERCENTILES

If you have not already done so generate a bell-shaped curve for SAT scores as described in Section 9.1. Make two copies of the plot so that you have three in all and use them for the examples below. Try to make your worksheet look like each example in turn.

Calculating Proportions

Example One
In the worksheet pictured next, the proportion of the area under the curve to the left of the 580 line has been calculated in cell G21. The command entered was =NORMDIST(580,500,100,1).

	G21	▼	=	=NORMDIST(580,500,100,1)				
	A	B	C	D	E	F	G	H	I
1	SAT score	height of curve							
2	200	4.43185E-05							
20	560	0.003332246							
21	580	0.002896916		proportion of scores below 580 is			0.78814467		
22	600	0.002419707							
23	620	0.001941861							
24	640	0.001497275							
25	660	0.001109208							
26	680	0.000789502							
27	700	0.00053991							
28	720	0.000354746							
29	740	0.000223945							
30	760	0.00013583							
31	780	7.91545E-05							
32	800	4.43185E-05							
33									
34									
35									
36									
37									
38									

The result, 0.78814467, means that approximately 79% of those tested scored below 580 on the SAT Exam. The percent scoring above 580 was 21%. ($100\% - 79\% = 21\%$.)

In general, the proportion of values for a normally distributed variable that fall below a particular value x can be found with the Excel command

$$=\text{NORMDIST}(x,\text{mean},\text{standard deviation},1)$$

Example Two

In the next example the proportion of students scoring between 450 and 650 is calculated by entering the command =NORMDIST(650,500,100,1)–NORMDIST(450,500,100,1) into cell G22.

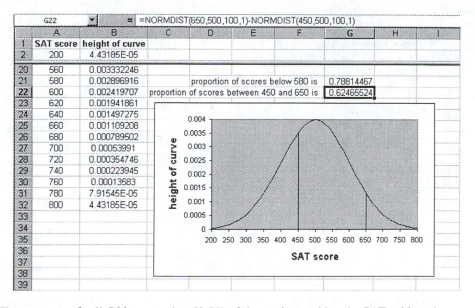

The contents of cell G22 means that 62.5% of the students taking the SAT achieved scores between 450 and 650.

In general, the proportion of values for a normally distributed variable that fall between values x and y, where $x < y$, can be found with the Excel command

=(NORMDIST(y,mean,standard deviation,1)

$$-\text{NORMDIST}(x,\text{mean},\text{standard deviation},1)$$

Percentiles

Example Three

In the next illustration, cell G23 gives the score below which 30% of the students performed. The command used was =NORMINV(0.30,500,100).

G23	▼	=	=NORMINV(0.3,500,100)					
	A	B	C	D	E	F	G	H	I

	A	B	C	D	E	F	G	H	I
1	SAT score	height of curve							
2	200	4.43185E-05							
20	560	0.003332246							
21	580	0.002896916		proportion of scores below 580 is			0.78814467		
22	600	0.002419707	proportion of scores between 450 and 650 is				0.62465524		
23	620	0.001941861			the 30th percentile is		447.5599		
24	640	0.001497275							
25	660	0.001109208							
26	680	0.000789502							
27	700	0.00053991							
28	720	0.000354746							
29	740	0.000223945							
30	760	0.00013583							
31	780	7.91545E-05							
32	800	4.43185E-05							
33									
34									
35									
36									
37									
38									

The number 447.56 in cell G23 means that a score of 448 (rounding to the nearest integer) on the SAT separates the lower 30% of all scores from the upper 70%. In other words, 448 is the 30th percentile.

The command NORMINV can be used to find a percentile p for any normal curve as follows:

$$=NORMINV(p, mean, standard\ deviation)$$

9.4 TESTING FOR NORMALITY

Drawing a Normal Curve Approximation to a Histogram

The techniques of this chapter can be applied only when it is reasonable to assume that the data in question has a bell-shaped or normal distribution. The proportions and percentiles calculated will not be correct otherwise. So before using these or any of the many other statistical computations that depend on the normality of the data, it is important to verify that the distribution of the data actually meets this criterion. A rough check of normality can be made by drawing a histogram of the data in question and including with it a normal curve approximation. If the curve fits the histogram fairly well, there is evidence of normality. The next picture shows a histogram of the ages of the world's 233 (as of 1992) billionaires.[1] The histogram is centered at about 65 years and is roughly symmetric with

---

[1]SOURCE: The Data and Story Library (DASL) web site. Reference: *Fortune*, Sept. 7, 1992, "The Billionaires," pp. 98-138.

tails extending to 7 years on the left and 102 or so years on the right. The outline of the histogram follows the normal curve approximation fairly well.[2]

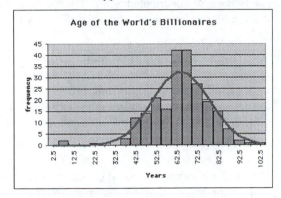

The next illustration shows a histogram of the wealth of these same billionaires. This is clearly a skewed distribution as more than half of the billionaires possess between 1 and 2 billion dollars. The others have wealth which spreads out far to the right. The normal curve approximation, as might be expected, is a poor fit.

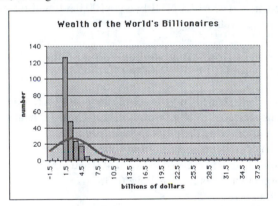

This means that although it might be reasonable to estimate the proportion of billionaires between the ages of, say, 35 and 70 using a normal curve and the techniques of this chapter, it would definitely be unreasonable to attempt to estimate the proportion whose wealth was, for example, greater than 2 billion dollars, using these same techniques.

Exercise 9.5 contains instructions for drawing normal curve approximations to histograms.

---

[2]The normal curve approximation drawn by the histogram add-in is not a density function as are those drawn in the previous sections. It is, instead, a normal frequency curve which is more convenient to use in this context as a quick geometric check of the normality of a distribution. The technical details of the distinction between the two types of curves need not be gone into at this point.

The 68%-95%-99.7% Rule

Data that is approximately normally distributed will follow the 68%-95%-99.7% Rule or, as it is sometimes called (depending on which statistics book you are reading), the *Empirical Rule*.

The 68%-95%-99.7% or Empirical Rule
68% of the data falls within 1 standard deviation of the mean. 95% of the data falls within 2 standard deviations of the mean. 99.7% of the data falls within 3 standard deviations of the mean.

This rule provides a numerical test of the normality of data. For example, the first criterion says that 68% of the data will lie within 1 standard deviation of the mean. This means that the mean minus 1 standard deviation should equal the 16th percentile, whereas the mean plus 1 standard deviation should equal the 86th percentile. The other two criteria provide similar tests. Exercise 9.8 contains detailed instructions for carrying out this test. Not surprisingly it makes use of Excel's PERCENTILE function. This function calculates, for a specified data range, the value below which any given percentage of the data lies. For example, the command =PERCENTILE(A1:A100,0.30) will return the number below which 30% of the entries in the range A1:A100 are located.

EXERCISES

Exercise 9.1 Draw a bell-shaped curve for SAT scores as described in Section 9.1. Make four copies of it, one for each of the exercises below. Draw vertical lines to set off the appropriate region and calculate the proportion asked for.
(a) What proportion of students scored below 600?
(b) What proportion of students scored between 400 and 700?
(c) What proportion of students scored above 650?
(d) Calculate the 70th percentile.
(e) Print the plot and the calculations.

Exercise 9.2 Young women aged 18 to 24 have an average height of 5 feet 4.3 inches with a standard deviation of 2.6 inches. For men in the same age range, the average height is 5 feet 10 inches with a standard deviation of 2.8 inches.
(a) Use Excel to draw a bell-shaped curve for each of these height distributions.
(b) What proportion of young women are taller than 5 feet 6 inches? young men? Draw vertical line(s) to set off the appropriate region.
(c) Calculate the percentile corresponding to your own height. Draw vertical line(s) to set off the appropriate region on the bell-shaped curve.
(d) Print the plot and the calculations.

Exercise 9.3 The Graduate Records Examination (GRE) has a mean of 497 and a standard deviation of 115 (ETS 1993). Draw a bell-shaped curve for this distribution. Suppose you are applying to a graduate school that requires a GRE score in the top 20%. How high must your score be to be eligible for admission? Draw a line on the curve that sets off the region of qualifying scores. Print the plot and the calculations.

The Standard Normal Curve

Exercise 9.4 The standard normal distribution has a mean of 0 and a standard deviation of 1. The command NORMSDIST(z) gives the percentile corresponding to a given z-value or standard score. The command NORMSINV(p) reverses this calculation and returns the standard score corresponding to the percentile value p.
(a) Draw a standard normal curve.
(b) Determine the percentage of a population falling between $z = -1$ and $z = 1$.
(c) Determine the z-score having 25% of the population above it.
(d) What z-score corresponds to the 20th percentile?
(e) Print the plot and the calculations.

Drawing a Normal Curve Approximation to a Histogram

Exercise 9.5 Open the worksheet **billionaires** in **datach9.xls**. Follow the instructions below to draw a normal curve approximation to the age data.
(a) Select the age data.
(b) Pull down the Tools menu and select Smart Histogram.
(c) In the Histogram Parameters dialog box select **Midpoint** under **Bin labels refer to** and under **Additional Options** check **Include Normal Curve Approximation**.
(d) Adjust the **Input Parameters** if you wish and click the OK button.
(e) Print the histogram.

Exercise 9.6 Open the worksheet **billionaires** in **datach9.xls**.
(a) Calculate the average and the standard deviation of the age data.
(b) Using the values found in part (a) as the mean and standard deviation of a normal distribution, estimate the proportion of billionaires 40 years of age or younger.
(c) How many billionaires are *actually* 40 years of age or younger? What proportion are they of the total? How does this number compare to the normal curve estimate?
(d) Print the calculations.

Exercise 9.7 Open the worksheet **billionaires** in **datach9.xls**.
(a) Draw a normal curve approximation to the wealth data.
(b) Calculate the average and the standard deviation of the data.
(c) Using the values found in part (b) as the mean and standard deviation of a normal distribution, estimate the proportion of billionaires whose wealth is $2 billion or more.
(d) How many billionaires are *actually* in this group? What proportion are they of the total? How does this number compare to the normal curve estimate?
(e) Print the plot and the calculations.

Exercise 9.8 Open the worksheet **billionaires** in **datach9.xls**. Follow the steps below to test the age data against the Empirical rule.

(a) Type table headings like those displayed next.

	A	B	C	D	E	F	G	H	I
1	age								
2	50								
3	88								
220	75								
221	62								
222	65				The Empirical Rule				
223	63				billionaire's age				
224	87		average minus 3sd			0.15 th	percentile		
225	61		average minus 2sd			2.5 th	percentile		
226	58		average minus 1sd			16 th	percentile		
227	60		average			50 th	percentile		
228	67		average plus 1sd			84 th	percentile		
229	80		average plus 2sd			97.5 th	percentile		
230	63		average plus 3sd			99.85 th	percentile		
231	9								
232	59								

(b) Enter the command =AVERAGE(A2:A232)–3*STDEV(A2:A232) into cell E224. The result should be 23.438. This means that the age 23.4 years is about 3 standard deviations below the average age of the billionaires.

(c) Enter the command =AVERAGE(A2:A232)–2*STDEV(A2:A232) into cell E225. The outcome is about 37 years. What does it mean?

(d) Complete the entry of commands into column E.

(e) Enter the command =PERCENTILE(A2:A232, 0.0015) into cell H224.

(f) Enter the command =PERCENTILE(A2:A232, 0.025) into cell H225. The outcome is 39.6, meaning that 2.5% of the ages fall below 39.6 years.

(g) Enter the command =PERCENTILE(A2:A232, 0.16) into cell H226. What does the resulting value of 52 mean?

(h) Continue to enter the appropriate percentile commands until column H has been calculated. When completed, the table should look like the worksheet pictured next. Print the table.

	A	B	C	D	E	F	G	H	I
1	age								
2	50								
3	88								
220	75								
221	62								
222	65				The Empirical Rule				
223	63				billionaire's age				
224	87		average minus 3sd		23.4385669	0.15 th	percentile	7.672	
225	61		average minus 2sd		36.969415	2.5 th	percentile	39.6	
226	58		average minus 1sd		50.5002631	16 th	percentile	52	
227	60		average		64.0311111	50 th	percentile	65	
228	67		average plus 1sd		77.5619592	84 th	percentile	77	
229	80		average plus 2sd		91.0928072	97.5 th	percentile	86.4	
230	63		average plus 3sd		104.623655	99.85 th	percentile	99.984	
231	9								
232	59								

Exercise 9.9 Follow the steps outlined in Exercise 9.8 to test the wealth data against the Empirical Rule.

Exercise 9.10 This exercise refers to the tables calculated in the previous two exercises. The Empirical Rule predicts agreement, or near agreement, between the percentiles in column H and the calculations in column E in the case of normally distributed data. Open a text box and write a comparison of the extent to which the two data sets, age and wealth, follow the Empirical Rule and print it.

WHAT TO HAND IN: Print no more than one page per exercise, showing the appropriate bell-shaped curves and Excel calculations. Position the curves and calculations to use space efficiently.

Investigating Sampling Distributions: The Central Limit Theorem

10

In This Chapter...

- The Sampling Tool
- Random sampling from a population
- Histograms of sample means

The Central Limit Theorem tells us that the distribution of sample means drawn from a population will be normally distributed if the population itself is normal. If the underlying population is not normal, then the distribution of sample means will nevertheless be approximately normal if the sample size n is large, say greater than 30. Further, if the population from which the samples are drawn has a mean μ and a standard deviation σ, then the average of the sample means will also be μ, whereas the standard deviation of the sample means will be approximately $\sigma/\sqrt{n}$, assuming n is small compared to the size of the population.[1] In this chapter you will take random samples from a variety of populations and investigate the distributions of the means of these samples, checking them against the claims of the Central Limit Theorem. The population data sets can be found in **datach10.xls**.

10.1 THE STUDENT HEIGHT DATA

Open the **heightdata** worksheet in the workbook **datach10.xls**. This data set contains the heights of 1000 college students. Copy this data into a new workbook, and, to get an idea of the distribution of the data, calculate its average, median, and standard deviation and draw a histogram for it. The result might look something like the next illustration.

---

[1] A sample size of 5% of the population size or less is usually considered to be sufficiently small. For larger sample sizes, the sample standard deviation is $\sigma/\sqrt{n}$ is multiplied by the correction factor $\sqrt{(N-n)/(N-1)}$. For this first investigation of the Central Limit theorem, these technicalities will not be dwelt on.

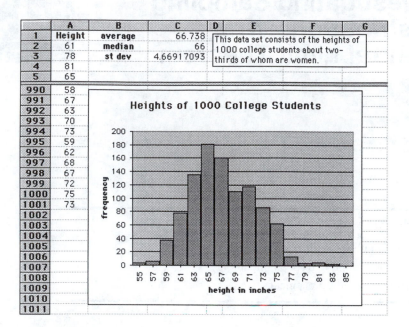

10.2 THE SAMPLING ADD-IN

In this section an Excel add-in called SRS (Simple Random Sample) will be used to select a specified number of samples of a given sample size from a range of data. The sampling process is *random* meaning that any element of the data range is just as likely to be selected for inclusion in the sample as any other. The selections can be made *with replacement*, so that the same data value can appear in a given sample more than once, or *without replacement*, meaning that once a data value is selected for a sample it is removed from the data pool and cannot be selected again. For the purposes the investigations of the workings of the Central Limit Theorem considered here, it does not matter which replacement option is picked.

Make Sure that Sampling Add-In is Available

The add-in is run by selecting SRS from the Tools menu. Pull down your Tools menu and make sure that SRS is listed. See the following illustration.

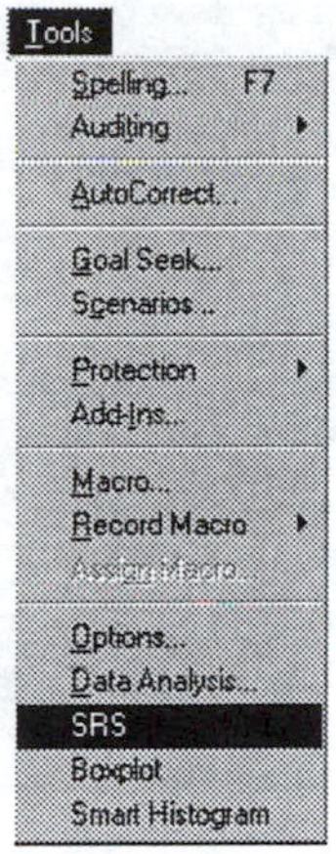

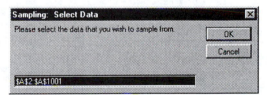

If SRS is nowhere on the Tools menu, consult Appendix B for instructions on installing it.

Using the Sampling Add-In

Use the sampling add-in, SRS, to generate 20 samples each of size 40 taken from the height data.

1. Select the height data. Click in cell A2, hold down the SHIFT key, and click in cell A1001.
2. Select **SRS** from the Tools menu. The Sampling: Select Data dialog box will appear. Make sure that the data range, A2:A1001, is typed in to the data box as shown next. (Alternatively, you can name this data range and refer to it by its name in this and subsequent calculations.)

3. Click the OK button.
4. When the Simple Random Samples dialog box appears, enter **20** as the **Number of Samples** and **40** as the **Sample Size**. The **Replacement?** option can be left at the default setting of **Yes**, as it is in the next illustration, or, if you prefer, at **No**.

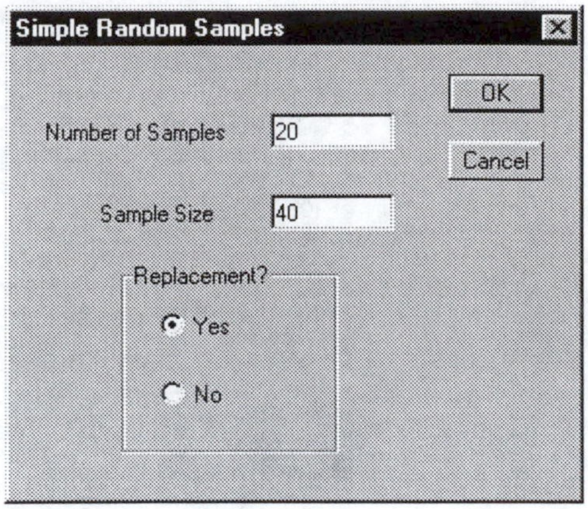

Simple Random Samples   ☒

Number of Samples   `20`

Sample Size   `40`

Replacement?
 ⦿ Yes
 ○ No

OK   Cancel

5. Click the OK button.

6. After, perhaps, some effort (the exact length of time required varies with the speed and power of the computer used), Excel will produce the 20 samples and place them in a numbered SRS sheet as shown next.

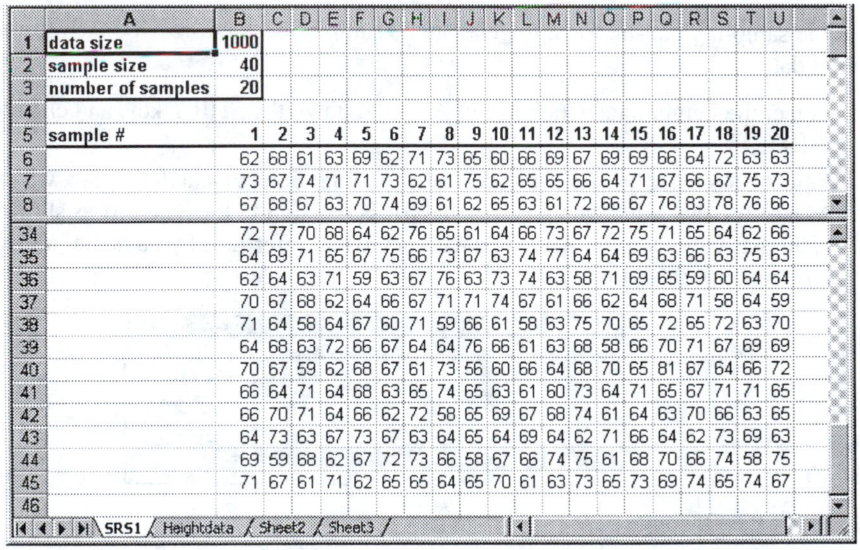

	A	B	C	D	E	F	G	H	I	J	K	L	M	N	O	P	Q	R	S	T	U
1	data size	1000																			
2	sample size	40																			
3	number of samples	20																			
4																					
5	sample #	1	2	3	4	5	6	7	8	9	10	11	12	13	14	15	16	17	18	19	20
6		62	68	61	63	69	62	71	73	65	60	66	69	67	69	69	66	64	72	63	63
7		73	67	74	71	71	73	62	61	75	62	65	65	66	64	71	67	66	67	75	73
8		67	68	67	63	70	74	69	61	62	65	63	61	72	66	67	76	83	78	76	66
34		72	77	70	68	64	62	76	65	61	64	66	73	67	72	75	71	65	64	62	66
35		64	69	71	65	67	75	66	73	67	63	74	77	64	64	69	63	66	63	75	63
36		62	64	63	71	59	63	67	76	63	73	74	63	58	71	69	65	59	60	64	64
37		70	67	68	62	64	66	67	71	71	74	67	61	66	62	64	68	71	58	64	59
38		71	64	58	64	67	60	71	59	66	61	58	63	75	70	65	72	65	72	63	70
39		64	68	63	72	66	67	64	64	76	66	61	63	68	58	66	70	71	67	69	69
40		70	67	59	62	68	67	61	73	56	60	66	64	68	70	65	81	67	64	66	72
41		66	64	71	64	68	63	65	74	65	63	61	60	73	64	71	65	67	71	71	65
42		66	70	71	64	66	62	72	58	65	69	67	68	74	61	64	63	70	66	63	65
43		64	73	63	67	73	67	63	64	65	64	69	64	62	71	66	64	62	73	69	63
44		69	59	68	62	67	72	73	66	58	67	66	74	75	61	68	70	66	74	58	75
45		71	67	61	71	62	65	65	64	65	70	61	63	73	65	73	69	74	65	74	67
46																					

SRS1 / Heightdata / Sheet2 / Sheet3 /

Since random processes are involved in generating the samples, your results will not be identical to the illustration. Split the screen as necessary so that rows 6 and 45 are both visible.

10.3 PICTURING THE DISTRIBUTION OF SAMPLE MEANS

Calculate the Sample Means

1. Type the row title "sample mean" into cell A46.
2. Enter the command =AVERAGE(B6:B45) into cell B46. A possible result is shown next.

B46		=	=AVERAGE(B6:B45)					
	A	B	C	D	E	F	G	H
1	data size	1000						
2	sample size	40						
3	number of samples	20						
4								
5	sample #	1	2	3	4	5	6	7
6		62	68	61	63	69	62	71
7		73	67	74	71	71	73	62
8		67	68	67	63	70	74	69
35		64	69	71	65	67	75	66
36		62	64	63	71	59	63	67
37		70	67	68	62	64	66	67
38		71	64	58	64	67	60	71
39		64	68	63	72	66	67	64
40		70	67	59	62	68	67	61
41		66	64	71	64	68	63	65
42		66	70	71	64	66	62	72
43		64	73	63	67	73	67	63
44		69	59	68	62	67	72	73
45		71	67	61	71	62	65	65
46	sample mean	66.2						
47								

This output says that the average height in the first sample of 40 height values is 66.2 inches.

3. Fill the average formula from cell B46 to cell U46: select cell B46, then hold down the SHIFT key and click cell U46. Pull down the Edit menu and select Fill and then Right.
4. Recalculate the average commands: hold down the SHIFT key and press the F9 key.
5. Widen columns C through U a bit so that several places are visible for each of the sample mean values. One possible result is pictured next.

	A	B	C	D	E	Q	R	S	T	U	V
1	data size	1000									
2	sample size	40									
3	number of samples	20									
4											
5	sample #	1	2	3	4	16	17	18	19	20	
6		62	68	61	63	66	64	72	63	63	
7		73	67	74	71	67	66	67	75	73	
8		67	68	67	63	76	83	78	76	66	
9		72	72	64	64	63	69	72	67	60	
35		64	69	71	65	63	66	63	75	63	
36		62	64	63	71	65	59	60	64	64	
37		70	67	68	62	68	71	58	64	59	
38		71	64	58	64	72	65	72	63	70	
39		64	68	63	72	70	71	67	69	69	
40		70	67	59	62	81	67	64	66	72	
41		66	64	71	64	65	67	71	71	65	
42		66	70	71	64	63	70	66	63	65	
43		64	73	63	67	64	62	73	69	63	
44		69	59	68	62	70	66	74	58	75	
45		71	67	61	71	69	74	65	74	67	
46	sample mean	66.2	66.4	66.3	66.1	67.3	66.9	67.5	68.7	65.7	
47											

These 20 sample mean values can now be treated as a data set having an average, a standard deviation, and a distribution that can be investigated with a histogram or boxplot.

Find the Average and Standard Deviation of the Sample Means

1. Type "average of sample means" into cell A48.
2. Enter the formula =AVERAGE(B46:U46) into cell B48.
3. Type "stand dev. of sample means" into cell A49.
4. Enter the formula =STDEV(B46:U46) into cell B49.
5. The result is shown next.

48	average of sample means	66.7
49	stand dev. of sample means	0.76

Construct a Histogram of the Sample Mean Values

1. Select Smart Histogram from the Tools menu.
2. When the Range dialog box appears, type B46:U46 into the Range box.
3. Make selections in the Parameters dialog box to produce a "nice" histogram of the 20 sample means. An example is shown next. It was copied from its histogram sheet and pasted into the SRS sheet.

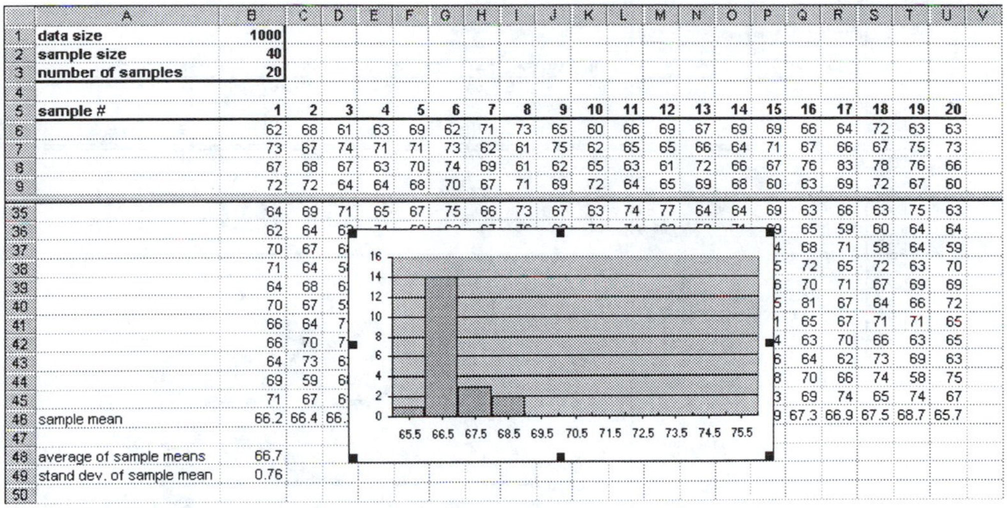

	A	B	C	D	E	F	G	H	I	J	K	L	M	N	O	P	Q	R	S	T	U	V
1	data size	1000																				
2	sample size	40																				
3	number of samples	20																				
4																						
5	sample #	1	2	3	4	5	6	7	8	9	10	11	12	13	14	15	16	17	18	19	20	
6		62	68	61	63	69	62	71	73	65	60	66	69	67	69	69	66	64	72	63	63	
7		73	67	74	71	71	73	62	61	75	62	65	65	66	64	71	67	66	67	75	73	
8		67	68	67	63	70	74	69	61	62	65	63	61	72	66	67	76	83	78	76	66	
9		72	72	64	64	68	70	67	71	69	72	64	65	69	68	60	63	69	72	67	60	
35		64	69	71	65	67	75	66	73	67	63	74	77	64	64	69	63	66	63	75	63	
36		62	64	62												69	65	59	60	64	64	
37		70	67	6												4	68	71	58	64	59	
38		71	64	5												5	72	65	72	63	70	
39		64	68	6												6	70	71	67	69	69	
40		70	67	5												5	81	67	64	66	72	
41		66	64	7												1	65	67	71	71	65	
42		66	70	7												4	63	70	66	63	65	
43		64	73	6												6	64	62	73	69	63	
44		69	59	6												8	70	66	74	58	75	
45		71	67	6												3	69	74	65	74	67	
46	sample mean	66.2	66.4	66.												9	67.3	66.9	67.5	68.7	65.7	
47																						
48	average of sample means	66.7																				
49	stand dev. of sample mean	0.76																				
50																						

10.4 REVISITING THE CENTRAL LIMIT THEOREM

The Central Limit Theorem tells us what to expect of a distribution of sample averages: it should be approximately bell-shaped, centered at the population mean with a standard deviation approximately equal to the population's standard deviation divided by the square root of the sample size. Even this very small set of just 20 sample means comes fairly close to these expectations. In our example, the average of the sample means is 66.7. The population average is 66.7. The standard deviation of the 20 sample means is 0.76, whereas the value predicted by the Central Limit theorem is $4.67/\sqrt{40}$ or 0.74. Not too bad. The histogram is a bit more problematic. When more samples are taken the bell-shaped character of the distribution of sample means is more obvious as, for example, in the following illustration for 250 samples.

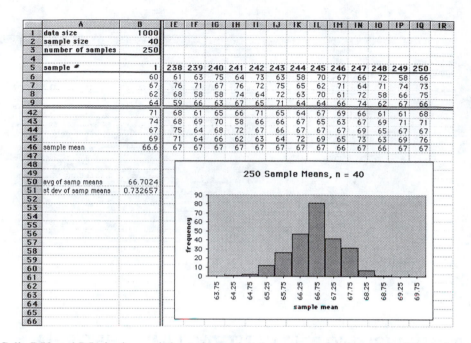

Cells B50 and B51 in the worksheet above give the average and the standard deviation, respectively, of the 250 sample means, and both values agree quite closely with the values predicted by the Central Limit Theorem.

EXERCISES

Exercise 10.1 Take 20 samples of size 40 from **heightdata** as described in Section 10.2. Compare the sample averages, standard deviation, and histogram to those predicted by the Central Limit Theorem. Print a worksheet containing these comparisons. Do *not* print the samples.

Exercise 10.2 Construct a worksheet like the one displayed at the end of Section 10.2 for 250 samples of size 40 taken from the height data. (If your machine balks or takes too long to generate the 250 samples, try generating the samples in batches. You can then combine the batches in a single worksheet. If all else fails, reduce the number of samples.) Print the histogram you obtain.

Exercise 10.3 Calculate the mean and the standard deviation of the sample means found in Exercise 10.2. How close do these values come to those predicted by the Central Limit Theorem? Print the calculations and discussion.

Exercise 10.4 Use the Empirical Rule to test the normality of the set of sample means you found in Exercise 10.2. Open a text box and write a brief discussion of your findings. Print the results of your test together with the text box.

Exercise 10.5 Repeat Exercises 10.2 and 10.3 using a much smaller sample size, say 5 or 10. Discuss any differences you observe when a smaller sample size is used in a text box and print it.

Exercise 10.6 Use the data in the worksheet **pop1** for this exercise. It consists of 800 numbers selected randomly from the interval between 0 and 1.
  (a) Construct a histogram for the data and compute its average and standard deviation. Write a brief description of the distribution.
  (b) Take 200 samples, each of size 30, from **pop1**.
  (c) Draw a histogram of the 200 sample means calculated in part (b). What is the shape of the distribution?
  (d) What are the average and standard deviation of the 100 sample means? How close are these values to those predicted by the Central Limit Theorem?
  (e) Does the Empirical Rule hold for the sample means?
  (f) Print the plot, the calculations, and the discussions.

Exercise 10.7 Use the data in the worksheet **dynamic sampling** for this exercise. It contains 5000 numbers selected randomly from the interval between 0 and 1. These values are contained in the range A2:J502. The mean value for each ten-number row was calculated in column K so that the range K2:K502 contains all of these means. Each of these means can be thought of as the average of a sample of size 10 drawn from the original 5000 value data set. A histogram of the data and of the sample means have already been drawn and should be visible in the top left corner of the worksheet. The range K506:K509 in the worksheet contains calculations relevant to the Central Limit Theorem.

This worksheet was set up so that you can alter the population at will to obtain a new set of random numbers, and when the population values change, Excel will automatically adjust all of the calculations that are linked to this data—the sample means, the histograms, and the Central Limit Theorem values. Try it: hold down the SHIFT key and press the F9 key, which will cause Excel to recalculate all of the functions in the worksheet, and watch the changes take place.
  (a) Hold down the SHIFT key and press the F9 key several times and observe the changes in the histograms of the population and of the sample means. Write a brief description of the two histograms and explain their relevance to the Central Limit Theorem.
  (b) Hold down the SHIFT key and press the F9 key 20 or 25 times, observing the values in the range K506:K509. What is the largest difference you observe in cell K509? To what extend are your observations consistent with the predictions of the Central Limit Theorem?

WHAT TO HAND IN: Print the histograms, text box discussions, and the mean and standard deviation calculations. Select the areas of your worksheets containing these entries carefully and be sure to pick **Print Selection** from the Print dialog box.

Exploring Probability

11

In This Chapter...

- RANDBETWEEN
- More about IF
- Coin tosses
- Relative frequency and empirical probability
- Charting relative frequency

Many ideas about probability can be explored using Excel. The next two sections discuss the commands and tools that are especially helpful in the simulation of random events. An important prerequisite for this work is the material in Section 8.1. If you are not familiar with this material, study it now before going on.

11.1 TWO EXCEL FUNCTIONS

RANDBETWEEN

This function is similar to the RAND command as it also selects a number at random, but the value it returns is an integer. For example, =RANDBETWEEN(1,10) returns an integer from the set 1,2,3,4,5,6,7,8,9,10. Follow the steps below to obtain a list of five randomly selected integers.

1. Open a worksheet. Make sure that Excel's Calculation option has been set at Manual. (See Section 8.1 for instructions.)
2. Enter the command =RANDBETWEEN(1,10) into cell A1.
3. Select cell A1. Move the cursor to the lower right corner of cell A1 and, when it becomes a black cross, press and drag to cell A5.
4. Hold down the SHIFT key and press the F9 key.

The result is a list of five integers selected at random from the set 1,2,3,4,5,6,7,8,9,10.

	=	=RANDBETWEEN(1,10)			
	A	B	C	D	E
1	7				
2	5				
3	4				
4	6				
5	10				
6					
7					

Remember, since each of these selections is made randomly, your list of integers will be a different set of values. This command will come up again in the exercises where it is used to simulate the roll of a die.

More About IF

The IF command is one of Excel's so-called logical functions. It was introduced earlier in Section 2.5 and is designed to make decisions concerning the contents of a cell. For example, enter the following command into cell B1 of the previous worksheet.

$$=IF(A1<7,1,0)$$

The result is shown next.

	=	=IF(A1<7,1,0)			
	A	B	C	D	E
1	7	0			
2	5				
3	4				
4	6				
5	10				

Why the 0 in cell B1? The IF command works as follows: it tells Excel to look in cell A1 and decide if cell A1 contains a number smaller than 7 or not. If the number is smaller than 7, Excel is to indicate this fact by printing a 1 in the active cell, otherwise, the number 0 is printed. Since the number 9 is *not* smaller than 7, Excel followed orders and placed a 0 in cell B1.

For a further illustration of the IF command, copy the contents of B1 down using a black cross drag, ending with row 5, or double-click on the lower right corner of cell B1 which will accomplish the same result. (If the Calculation option is set at Manual, press SHIFT-F9 after copying in order to calculate all of the IF commands.)

	=	=IF(A5<7,1,0)			
	A	B	C	D	E
1	7	0			
2	5	1			
3	4	1			
4	6	1			
5	10	0			
6					

Notice that Excel placed the number 1 next to each number less than 7 and a 0 next to the numbers not less than 7. You will see more examples of the IF function later in the chapter.

11.2 SIMULATING TOSSES OF A FAIR COIN

Begin by making sure that Excel's **Calculation** option has been set at **Manual**. (See Section 8.1 for instructions.) Open a fresh worksheet.

One Toss

The task is to design an Excel command that will behave like a fair coin. When a fair coin is tossed, it lands with either heads or tails on top. One side is just as likely to appear as the other. To model this phenomenon we need a command that will return two different values, but in such a way that both are equally likely. It must be designed so that we cannot predict ahead of time which of the two values will be returned, just as we don't know until we have flipped it how a coin will land. Study the command displayed below. It does exactly what is required.

$$=IF(RAND()<0.5,1,0)$$

In carrying out this command, Excel will first execute the function RAND(), which will generate a random number between 0 and 1. Excel does not print this number. Instead, it keeps the random number in memory and compares it with 0.5. If the random number is less than 0.5, it prints a 1 in the active cell. If the random number is not less than 0.5, meaning it is between 0.5 and 1, Excel prints a 0. Enter the command into your worksheet. One possible result is shown next.

Imagine a coin with the number 1 embossed on one side and the number 0 on the other. The command models the act of flipping such a coin. In the illustration above, the coin landed with the number 1 showing. Now, when we interpret the 1 as heads and the 0 as tails, the command becomes a model for the toss of an ordinary fair coin. Hold down the SHIFT key and press the F9 key repeatedly. This action is equivalent to tossing a fair coin over and over again—when a 1 appears, think "heads" and when a 0 appears think "tails."

With this little discussion behind us we are ready to model 10 tosses of a fair coin. Just follow these steps:

Ten Tosses

1. Make sure that **Calculation** is set to **Manual**. (See Section 8.1.)
2. Enter the command =IF(RAND()<0.5,1,0) into cell A1 of a worksheet.

3. Hold down the SHIFT key and click in cell A10. From the Edit menu select Fill Down.

4. Hold down the SHIFT key and press the F9 key. The ten IF commands in A1:A10 will be recalculated to produce a list of ten 0's and 1's. A possible outcome is displayed next.

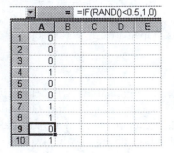

5. Click in cell A11 and press the Sum button on the toolbar. Press the ENTER key and Excel will return the number of "heads" out of the 10 tosses.

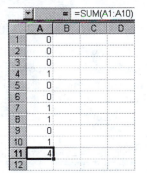

6. Hold down the SHIFT key and press the F9 key repeatedly. Observe how the total number of heads changes. Most values are near 5, but occasionally numbers as small as 2 or as large as 8 are recorded. Much, much more rarely, you will see 9 or 10 heads or, on the low end, 0 or 1. These totals are what would be expected with 10 tosses of a real coin.

11.3 SIMULATING RELATIVE FREQUENCY

One of the common interpretations of probability is as a measure of what will happen when an event, such as flipping a coin, is repeated over and over again. The probability of the event is the proportion of times the event occurs in the long run. This proportion is called the *relative frequency* of the event. Your statistics textbook will have a great deal more to say about this subject. Our purpose here is to give an example of an Excel simulation of relative frequency.

Begin with a quick hand calculation to review the idea of relative frequency. If a coin is tossed four times with the outcome HHHT, then the relative frequency of heads is 3/4 or 0.75. In general the relative frequency of heads in repeated tosses of a coin is calculated by dividing the total number of heads by the total number of tosses. When coin tosses are simulated in Excel, as described in the previous section, the relative frequency of heads can be quickly calculated using the AVERAGE command. The next illustration shows an example for the simulated outcome HHHT.

=AVERAGE(A1:A4)

	A	B	C	D	E
1	1				
2	1				
3	0				
4	1	0.75			
5					

The AVERAGE command works here because when it is applied to the range A1:A4 it divides the sum of the range values, 3, by the number of entries in the range, 4. The sum of the range is the total number of heads, whereas the number of values in the range is the total number of tosses. Hence, average in this case *is* relative frequency.

The next project is to keep track of the changes in relative frequency as the number of tosses increases. In other words, to see what happens to the relative frequency of heads in the long run. The following exercises are designed to help you with this investigation.

EXERCISES

Exercise 11.1 Begin by simulating 100 tosses of a fair coin together with a running calculation of the relative frequency of heads. To do so follow the procedure outlined below.

(a) Open a new worksheet and make sure that **Calculation** is set at **Manual**. (See Section 8.1.)

(b) Split the screen horizontally so that rows 1 and 103 are both visible.

(c) Enter column headings such as the ones shown below.

	A	B	C
1			**Relative**
2	**toss**	**H = 1, T = 0**	**Frequency**
3			
4			
5			
6			

(d) Number the tosses in column A as follows: Enter the number 1 into cell A3. Click on cell A3. Hold down the SHIFT key and click on cell A102. Pull down the Edit menu select Fill and then Series... . The Series dialog box will appear. Make sure that it is filled in to match the next illustration.

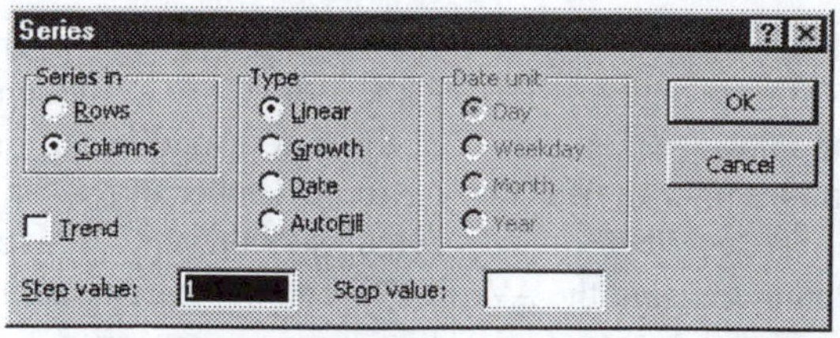

Click the OK button. The numbers 1 through 100 should now appear in range A3:A102.

(e) Enter the command =IF(RAND()<0.5,1,0) into cell B3.

(f) Enter the command =AVERAGE($B$3:$B3) into cell C3. (Make sure that you type in the $-signs just as they are written. Otherwise the command will not work when it is filled down.) Your worksheet should now have either the number 0 or the number 1 in cells B3 and C3 as in the next illustration.

= =AVERAGE($B$3:$B3)

	A	B	C	D
1			Relative	
2	toss	H = 1, T = 0	Frequency	
3	1	1	1	
4	2			
5	3			
6	4			
7	5			
99	97			
100	98			
101	99			
102	100			

(g) Select the range B3:C102: click on cell B3, then hold down the SHIFT key and click on cell C102.

(h) Pull down the Edit menu and select Fill and then Down.

(i) Recalculate the 100 IF and AVERAGE commands in columns B and C: Hold down the SHIFT key and press F9. The result will look something like the next illustration.

= =AVERAGE($B$3:$B102)

	A	B	C	D
1			Relative	
2	toss	H = 1, T = 0	Frequency	
3	1	0	0	
4	2	1	0.5	
5	3	0	0.333333333	
6	4	0	0.25	
7	5	1	0.4	
99	97	0	0.474226804	
100	98	1	0.479591837	
101	99	0	0.474747475	
102	100	0	0.47	
103				

When you have completed these steps, the range B3:C102 will contain a simulation of 100 tosses of a fair coin in column B together with the running relative frequency of heads in column C. The next step is to get an idea of how these relative frequencies behave as the number of tosses increases. A picture will tell the tale.

Charting Relative Frequency in Excel 5 or 7

The next set of instructions illustrate how you can use Chart Wizard to plot the running relative frequencies. Once this picture is drawn it will be easier to understand the long-run behavior of the coin tosses.

(a) Select the relative frequency data: click in cell C3, hold down the SHIFT key and click in cell C102.
(b) Click the Chart Wizard button. At Step 1 verify that the range is correct.
(c) Select the **Line** plot at Step 2 and Option **2** of the line plot at Step 3.
(d) At Step 4 press the Next button.
(e) At Step 5 remove the legend and label the chart and the axes. Click on the Finish button. Your chart should look something like the one shown next.

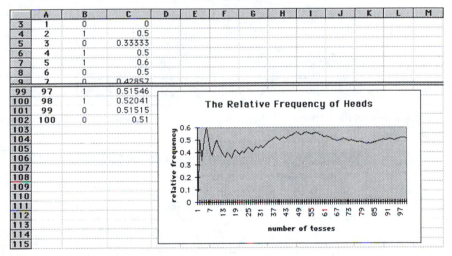

Don't worry if your line plot has peaks and valleys at different places or if the axes are labeled somewhat differently. Just make sure that it is a plain line graph containing no enlarged points. The next several steps outline instructions for making the graph more readable.

Charting Relative Frequency in Excel 97

1. Launch Chart Wizard and at Step 1 select the **Line** chart type the sub-type appearing at the top left.
2. At Step 2 click the Next button.
3. At Step 3 remove the legend and type in titles.
4. At Step 4 click the Next button.

Editing a Relative Frequency Chart

1. Activate the relative frequency chart.
2. The horizontal axis has too many tick marks. Double-click on the horizontal axis to open the Format Axis dialog box. Click on the **Scale** tab to enter the values displayed in the next illustration.

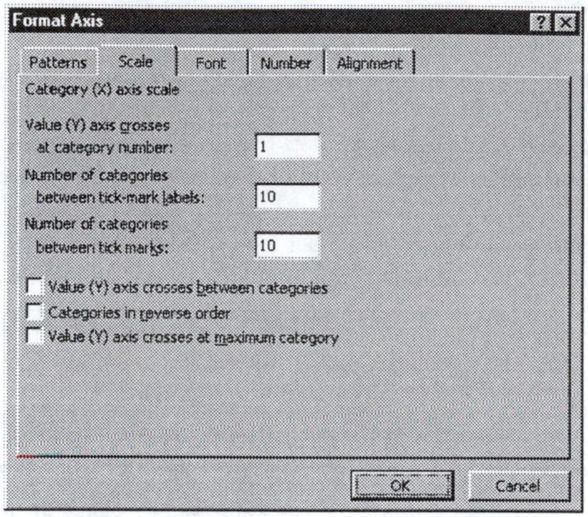

3. Double-click on the vertical axis to reopen the Format Axis dialog box. Click on the **Scale** tab and enter the values displayed in the next illustration.

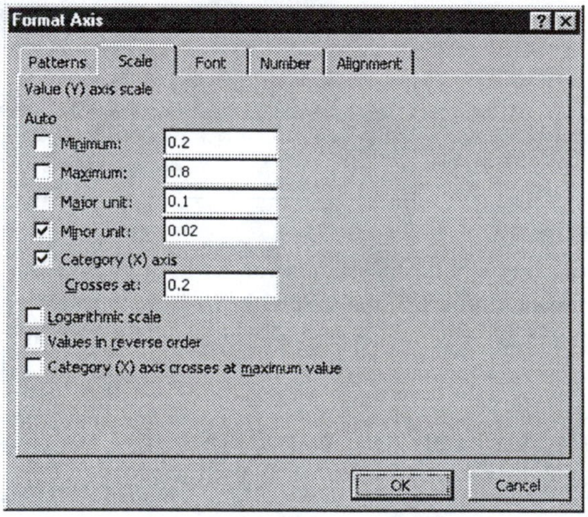

4. If the chart has no horizontal gridlines, insert them. In Excel 5 or 7 pull down the Insert menu and select Gridlines... . In Excel 97 select Chart Options from the Chart menu. In either case select the option that draws **Major gridlines** for the **Value (Y) Axis**.

You should now have a beautiful relative frequency chart looking something like the one shown next.

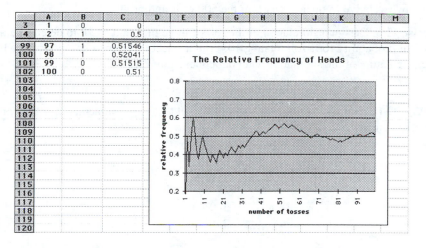

	A	B	C	D	E	F	G	H	I	J	K	L	M
3	1	0	0										
4	2	1	0.5										
99	97	1	0.51546										
100	98	1	0.52041										
101	99	0	0.51515										
102	100	0	0.51										
103													
104													
105													
106													
107													
108													
109													
110													
111													
112													
113													
114													
115													
116													
117													
118													
119													
120													

Open a text box and write a brief discussion of the behavior of your frequency line. Print a copy of your chart together with the discussion.

Exercise 11.2 You can recalculate the 100 tosses you simulated in the previous exercise by holding down the SHIFT key and pressing F9. Do 100 recalculations, keeping track of the number of times that the relative frequency at the last toss lies below 0.4 or above 0.6. (Suggestion: Make a tick mark on a sheet of paper each time you observe such a frequency.) What percentage of the recalculations gave such a result? Print your calculations.

Exercise 11.3 Open the worksheet **500 tosses** in workbook **datach11.xls**. It contains a relative frequency chart for 500 tosses of a fair coin.
(a) Recalculate this worksheet several times. What differences do you notice between the 500 toss charts and the 100 toss charts you looked at in the previous exercises?
(b) Recalibrate the vertical axis so that it has a minimum of 0.45 and a maximum of 0.55. Do 50 or so recalculations of this worksheet and keep track of the number of times that the final relative frequency exceeds 0.53 or is smaller than 0.47. What percentage of the time do you observe such a result? Print your calculations.

Exercise 11.4 Simulate 100 tosses of an unfair coin for which the probability of heads is 0.7. Print a relative frequency chart for your simulation along with a text box discussion of your result.

Exercise 11.5 Open the worksheet **two dice** in **datach11.xls**. It contains a simulation of 500 rolls of a pair of dice. Answer parts (a) through (c) in a text box and print it.
(a) What commands appear in columns A and B? Explain why these commands simulate the roll of a die. Why do the commands in column C simulate the roll of a pair of dice?
(b) What is the purpose of the commands in column D?

(c) The number in cell E502 is an empirically determined probability. What command produced the number and what probability does it approximate?

(d) Draw a relative frequency chart for the 500 rolls and print it.

Exercise 11.6 Design a simulation of the roll of three dice and use it to determine an empirical probability for a roll totaling 10. Print a selection from your worksheet which makes the nature of the simulation and your conclusions clear.

WHAT TO HAND IN: Print just those charts and text box commentaries asked for. One page per exercise should suffice.

Election Polls: An Exploration of Sample Proportions

12

In This Chapter...

- Simulating an election poll
- Investigating the distribution of sample proportions
- Picturing confidence intervals
- Using $p = 0.5$ to estimate sample standard deviations

In a presidential preference poll published in early November 1996, 51% of those surveyed said they intended to vote for President Clinton, 38% for Robert Dole, the Republican candidate, and the remainder were undecided. The sample size was 1600.

We are all familiar with the results of such political polls. The pollster claims that a certain proportion of the electorate favors a candidate. A margin of error is usually given and sometimes a measure of confidence, such as 95%, will be reported. In this chapter we will look at the statistical underpinnings for election polling. We will employ Excel's RAND and IF commands to model the process of random selection. The RAND command was introduced in Section 8.1 and IF is discussed in Section 2.5 and again in Section 11.1. You might want to review these sections before going on.

12.1 A SIMULATED SAMPLE

Suppose that 60% of the residents of a certain state actually support a candidate named Hiddy Cole. In real life this proportion would, of course, be unknown. Suppose that Hiddy's campaign team commissions a poll to estimate the level of her support. The polling firm decides to use a random sample of size 300 to make this estimate. (In an actual poll this figure would be much higher, but for the purposes of this illustrative example $n = 300$ will do.) Since the sample is random, the probability is 0.60 that an individual included in the poll will be a Cole supporter. The Excel command

$$=IF((RAND() < 0.6, 1, 0)$$

where 1 corresponds to the selection of a Cole supporter and 0 to the selection of a supporter of one of the other candidates, can be used to simulate the political opinion of a randomly selected voter. By executing this command over and over again the process of sampling a large number of voters can be modeled.

Simulating a Sample of Size 300

Follow these instructions to simulate a random sample of 300 voters chosen from a population where 60% of the population support our hypothetical candidate, Hiddy Cole.

1.   Set the **Calculation** option to **Manual**. (See Section 8.1.)
2.   Open a worksheet and type the command =IF(RAND()<0.6,1,0) into cell B3 and give the worksheet an appropriate heading as shown next in row 1.

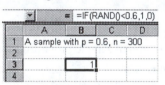

3.   Split the screen horizontally. Click in the lower portion of the divided screen and hit the PAGE DOWN key until row 302 is visible. Make sure that row 1 is still at the top of the upper screen. (Slide the upper scroll bar to the top of the upper screen.)
4.   Copy the IF command down to cell B302 as follows: Click in cell B3, hold down the SHIFT key and click in cell B302. Pull down the Edit menu and select Fill and then Down.
5.   Hold down the SHIFT key and press F9 in order to recalculate all of the IF commands. The result will be a column of 300 0's and 1's as shown next.

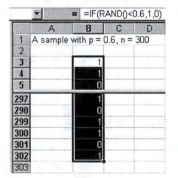

6.   The next step is to find the proportion of "persons" in the sample who are Hiddy Cole supporters. Recall that each 1 appearing in column B corresponds to a supporter; nonsupporters were registered as 0's. Thus, the sum of the column entries will give the total number of supporters in the sample. This total must be divided by the sample size in order to give the proportion. Enter the command =SUM(B3:B302)/COUNT(B3:B302) (or, equivalently, =AVERAGE(B3:B302)) into

cell B303 to obtain the proportion. In this case, the sample proportion is 0.56 as the next picture shows.

	=	=SUM(B3:B302)/COUNT(B3:B302)			
	A	**B**	**C**	**D**	**E**
1	A sample with p = 0.6, n = 300				
2					
3		1			
4		0			
5		0			
297		1			
298		0			
299		1			
300		1			
301		1			
302		0			
303	samp prop	0.56			
304					

7. Label row 303 as shown above with the heading "samp prop" for sample proportion.

The sample proportion, $\hat{p} = 56\%$, differs from the actual population proportion $p = 60\%$ by 4 percentage points. Some difference should be anticipated. It is not reasonable to expect the sample proportion to exactly duplicate the population proportion. However, if the sample size is large enough, most sample proportions will be near the population proportion. In the following exercises, you will investigate the nature of the distribution of such polling proportions.

EXERCISES

Generating a List of Sample Proportions

Exercise 12.1 Follow the steps below to generate 75 samples of 300 persons each and to calculate the proportion of Hiddy Cole supporters in each sample.
(a) Begin by generating one sample in range B3:B303 by the method described in the chapter.
(b) Split the screen horizontally and vertically so that rows 1 through 303 and columns A through BX are visible.
(c) Click in cell B3 and hold down the SHIFT key and click in cell BX303.
(d) Pull down the Edit menu and select Fill Right.
(e) Hold down the SHIFT key and press the F9 key so that all of the commands are calculated. An example of a possible result is shown next.

	A	B	C	D	E	F	G	H	I	J	K	BY	BW	BX
1	75 Sampling Simulations with n = 300, p = 0.6													
2														
3		1	0	0	1	0	1	1	0	1	1	0	1	0
4		1	1	0	1	1	0	1	1	0	1	1	0	0
5		1	1	1	1	1	0	1	1	1	0	1	1	0
297		1	0	1	1	1	1	1	1	1	1	1	0	1
298		0	1	1	1	1	1	0	1	1	0	0	0	0
299		1	1	1	1	0	1	1	0	1	1	0	0	0
300		0	1	1	1	0	1	1	1	1	1	1	1	0
301		1	1	1	0	1	1	1	0	0	1	1	1	1
302		0	0	0	1	1	0	1	1	1	0	1	1	1
303	sample prop	0.63	0.58	0.64	0.57	0.58	0.58	0.6	0.6	0.6	0.57	0.58	0.6	0.55

Notice that a border was added between the samples and the proportions.

Not enough memory?

Your machine may complain that it does not have enough memory to complete the right fill as it is outlined above. In this case, execute the right fill for a smaller number of columns, say 15, and repeat the right fill until 75 samples have been generated. If there are still memory problems, try closing any other open notebooks or applications and try again. As a last resort, reduce the number of samples and/or the sample size.

The row of sample proportions, row 303, can now be viewed as a data set. Its distribution will approximate that of all the possible samples of size 300 that might be drawn from the population. In the following exercises you will have a chance to study the shape and the mean and standard deviation of this distribution.

Determining the Shape of the Distribution of Sample Proportions

Exercise 12.2 Embed a histogram of the sample proportions, like the one illustrated next, into the sample proportions worksheet.

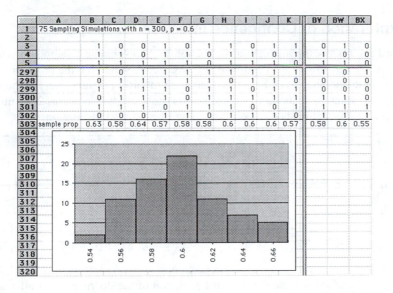

(a) Select the 75 sample proportions: click in cell B303, hold down the SHIFT key and click cell BX303.

(b) Select Smart Histogram from the Tools menu. Experiment until you get a nice chart for the proportions.

(c) When the histogram is drawn, copy it to the sheet containing the 75 samples. If you have forgotten how to do this, a reminder follows.

  (i) Click on the tab for the histogram sheet.

  (ii) Click on the histogram chart to select it.

  (iii)Pull down the Edit menu and select Copy.

  (iv)Click on the tab for the sheet containing the samples and then in cell B305.

  (v) Pull down the Edit menu and select Paste.

The result should look something like the worksheet shown at the beginning of this exercise.

(d) Press the F9 key. Excel will recalculate the 75 samples and redraw the histogram for the new data. Use F9 for this recalculation, *not* SHIFT-F9, since both the sample worksheet and the histogram worksheet must be recalculated. Remember, SHIFT-F9 will recalculate only the active worksheet, whereas F9 recalculates every worksheet.

Repeat the F9 recalculation several times, noting the center and shape of each histogram. Open a text box and write a description of your observations. Print one of the histograms together with the text box.

Exercise 12.3 Sampling theory tells us that the distribution of the sample proportions favoring a candidate will have a mean equal to p, the proportion of the population who actually favor the candidate, and a standard deviation equal to $\sqrt{\frac{p(1-p)}{n}}$, where n is the sample size. Calculate the mean and standard deviation of the 75 samples calculated in the previous exercise. How close do these values come to those predicted by the theory?

An Exploration of Confidence Intervals for Proportions

Exercise 12.4 Follow the instructions in Section 12.1 to simulate a sample of size 200 taken to assess support for the candidate Hiddy Cole. Assume that the proportion of supporters in the population is 60%. Calculate the proportion of supporters in the sample.

(a) Copy this sample simulation 19 times so that the worksheet contains 20 simulations altogether. The result should look something like the next illustration.

	A	B	C	J	K	L	M	N	O	P	Q	R	S	T	U
1				The Samples											
2		1	1	0	1	0	1	1	1	0	1	1	1	1	0
3		0	1	1	0	1	0	1	1	0	0	1	1	1	1
197		0	1	0	1	1	1	0	1	1	0	1	1	1	0
198		1	0	1	1	0	1	0	1	1	1	1	0	0	0
199		1	1	1	1	0	1	1	0	1	1	0	1	1	0
200		0	1	1	0	1	1	1	0	1	0	1	0	1	0
201		0	1	1	0	1	0	1	1	1	0	0	1	1	1
202	sample prop	0.61	0.62	0.60	0.55	0.65	0.59	0.56	0.58	0.61	0.58	0.63	0.62	0.59	0.59

(b) In the illustration, the first sample proportion, $\hat{p} = 0.61$, is located in cell B202. Use this value to estimate the standard deviation of the distribution of all sample proportions by entering the formula

$$=\text{SQRT(B202*(1–B202)/COUNT(B2:B201))}$$

into cell B203. In this example the standard deviation is 0.03.

B203		▼	=	=SQRT(B$202*(1-B$202)/COUNT(B2:B201))											
	A	B	C	J	K	L	M	N	O	P	Q	R	S	T	U
1				The Samples											
2		1	1	0	1	0	1	1	1	0	1	1	1	1	0
3		0	1	1	0	1	0	1	1	0	0	1	1	1	1
197		0	1	0	1	1	1	0	1	1	0	1	1	1	0
198		1	0	1	1	0	1	0	1	1	1	1	0	0	0
199		1	1	1	1	0	1	1	0	1	1	0	1	1	0
200		0	1	1	0	1	1	1	0	1	0	1	0	1	0
201		0	1	1	0	1	0	1	1	1	0	0	1	1	1
202	sample prop	0.61	0.62	0.60	0.55	0.65	0.59	0.56	0.58	0.61	0.58	0.63	0.62	0.59	0.59
203	standard deviation	0.03													
204															

(c) Fill the formula to the right so that a population standard deviation estimate is calculated from each sample proportion. The values are pictured below. Remember to press SHIFT-F9 to recalculate the formulas.

	A	B	C	J	K	L	M	N	O	P	Q	R	S	T	U
1				The Samples											
2		1	1	1	0	0	1	1	0	0	1	1	0	1	1
3		0	0	0	0	0	1	0	1	0	1	0	0	1	1
197		1	1	1	1	1	0	1	0	0	1	1	1	1	1
198		0	1	1	0	1	0	0	0	1	1	0	1	0	1
199		0	0	0	1	0	1	0	1	0	1	0	1	0	1
200		1	1	0	1	1	1	1	0	1	0	1	1	0	0
201		1	1	0	0	1	1	0	1	0	0	0	1	0	1
202	sample prop	0.60	0.63	0.57	0.54	0.61	0.63	0.66	0.57	0.59	0.62	0.59	0.59	0.58	0.58
203	standard deviation	0.03	0.03	0.04	0.04	0.03	0.03	0.03	0.04	0.03	0.03	0.03	0.03	0.03	0.03

(d) Calculate the 95% confidence interval for the population proportion using the first sample proportion and standard deviation. To find the lower limit enter the formula =B$202–1.96*B$203 into cell B206. (Remember to include the $ signs so that the formula will work when filled to the right.)

		=	=B$202-1.96*B$203	
	A		B	C
1				
2			1	1
3			0	0
197			1	1
198			0	1
199			0	0
200			1	1
201			1	1
202	sample prop		0.60	0.63
203	standard deviation		0.03	0.03
204	confidence interval			
205	for population prop			
206	lower bound		0.53	
207	upper bound			

(e) Now, type the formula for the upper limit, =B$202+1.96*B$203 into cell B207.

		=	=B$202+1.96*B$203	
	A		B	C
1				
2			1	1
3			0	0
197			1	1
198			0	1
199			0	0
200			1	1
201			1	1
202	sample prop		0.60	0.63
203	standard deviation		0.03	0.03
204	confidence interval			
205	for population prop			
206	lower bound		0.53	
207	upper bound		0.66	
208				

The result produces an estimate for the actual proportion of voters who support Ms. Cole, namely, $0.53 < p < 0.66$. This is the 95% confidence interval determined from the data in the first sample. Each of the other samples will also produce a confidence interval.

(f) Calculate these intervals. Click in cell B206. Hold down the SHIFT key and click in cell U207. Pull down the Edit menu and select Fill Right. Press SHIFT-F9 to recalculate all of the worksheet. This recalculation will apply to the whole worksheet so new samples will appear. The confidence intervals will, of course, be found for these new samples.

	A	B	C	J	K	L	M	N	O	P	Q	R	S	T	U
1				The Samples											
2		1	0	1	1	1	1	0	0	1	1	0	0	1	1
3		1	1	0	0	1	0	1	1	1	1	0	1	0	1
197		0	1	1	1	0	1	0	1	1	1	0	1	1	1
198		1	1	1	0	1	1	1	1	1	1	1	0	0	1
199		1	1	0	1	1	0	1	0	1	1	0	0	0	1
200		0	1	1	0	1	1	1	0	0	1	1	1	0	1
201		0	1	0	0	1	1	0	1	0	1	1	1	0	1
202	sample prop	0.59	0.61	0.55	0.63	0.57	0.58	0.61	0.56	0.67	0.69	0.6	0.57	0.57	0.62
203	standard deviation	0.03	0.03	0.04	0.03	0.04	0.03	0.03	0.04	0.03	0.03	0.03	0.04	0.04	0.03
204	confidence interval														
205	for population prop														
206	lower bound	0.52	0.54	0.48	0.56	0.5	0.51	0.54	0.49	0.6	0.62	0.53	0.5	0.5	0.55
207	upper bound	0.66	0.67	0.61	0.69	0.64	0.64	0.67	0.63	0.74	0.75	0.66	0.64	0.63	0.68

Notice that the confidence interval contained in column Q of the worksheet pictured above, $0.62 < p < 0.75$, does not include the population proportion. All of the other visible intervals do include the population proportion.

(g) How many of *your* confidence intervals contain the actual population proportion $p = 0.60$? Print rows 206 and 207 of your worksheet.

Picturing Confidence Intervals

Exercise 12.5 A picture is worth a thousand words so here are instructions for charting confidence intervals.

In Excel 5 and 7

(a) Generate 20 samples and construct 95% confidence intervals for each using the method described in the previous exercise.

(b) Click in cell B206. Hold down the SHIFT key and click in cell U207.

(c) Click the Chart Wizard button. Select the **Line** option at Step 2 and option **7** at Step 3.

(d) Keep the legend. Give the chart and the axes appropriate titles. The result will look something like the chart pictured next.

	A	B	J	K	L	M	N	O	P	Q	R	S	T	U	V	W
1			The Samples													
2		0	1	1	1	1	0	1	1	0	1	1	0	0		
3		1	0	1	1	1	1	1	0	1	0	1	0	1		
202	sample prop	0.62	0.62	0.63	0.58	0.57	0.62	0.62	0.61	0.56	0.57	0.63	0.65	0.59		
203	standard deviation	0.03	0.03	0.03	0.03	0.04	0.03	0.03	0.03	0.04	0.04	0.03	0.03	0.03		
204	confidence interval															
205	for population prop															
206	lower bound	0.55	0.55	0.56	0.51	0.5	0.55	0.55	0.54	0.49	0.5	0.56	0.58	0.52		
207	upper bound	0.69	0.69	0.69	0.64	0.64	0.69	0.69	0.68	0.63	0.64	0.69	0.72	0.66		

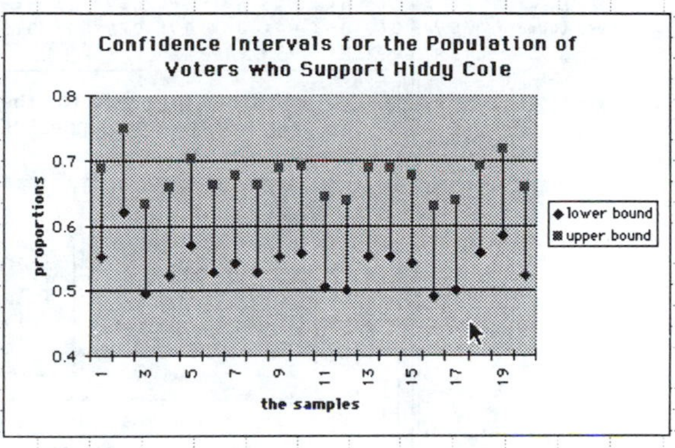

Confidence Intervals for the Population of Voters who Support Hiddy Cole

Notice that the vertical axis has been rescaled and gridlines were added.

Using the picture it is easy to see how many of the confidence intervals miss the population proportion. In this illustration, only the confidence interval constructed from the second sample misses the population proportion. The remaining 19 include the population proportion.

In Excel 97

(a) Generate a group of 20 samples and the upper and lower limits of their 95% confidence intervals as described in Exercise 12.5.

(b) Calculate the sample proportions in row 208.
   (i) Click in cell B208 and enter the formula =SUM(B2:B201)/COUNT(B2:B201).
   (ii) Click in cell B208, hold down the SHIFT key, and click in cell U208.
   (iii) Select Fill and then Right from the Edit menu.
   (iv) Recalculate with SHIFT-F9.

(c) Click in cell A206. Hold down the SHIFT key and click in cell U208.

(d) Click the Chart Wizard button. Select the **Stock** chart type at Step 1 and the default sub-type, which is located at the left of the top row of options.

(e) Remove the legend. Give the chart and the axes appropriate titles. The result will look something like the chart pictured next.

	A	B	J	K	L	M	N	O	P	Q	R	S	T	U	V	W
1			The Samples													
2		1	1	1	1	0	1	1	1	1	0	1	0	1		
3		1	1	0	1	0	0	1	1	1	1	0	1	1		
4		0	0	0	1	0	1	1	1	0	0	1	1			
201		0	1	1	0	0	1	1	1	0	0	1	1	1		
202	sample prop	0.67	0.59	0.61	0.64	0.58	0.64	0.61	0.67	0.59	0.6	0.59	0.63	0.61		
203	standard deviation	0.03	0.03	0.03	0.03	0.03	0.03	0.03	0.03	0.03	0.03	0.03	0.03	0.03		
204	confidence interval															
205	for population prop															
206	lower bound	0.6	0.52	0.54	0.57	0.51	0.57	0.54	0.6	0.52	0.53	0.52	0.56	0.54		
207	upper bound	0.74	0.65	0.68	0.71	0.64	0.71	0.68	0.74	0.65	0.66	0.65	0.7	0.67		
208		0.67	0.59	0.61	0.64	0.58	0.64	0.61	0.67	0.59	0.6	0.59	0.63	0.61		

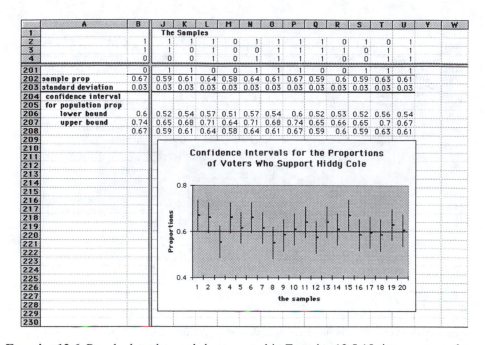

Confidence Intervals for the Proportions of Voters Who Support Hiddy Cole

Exercise 12.6 Recalculate the worksheet created in Exercise 12.5 10 times or more by repeatedly pressing the F9 key. Keep track of the number of confidence intervals that do *not* include the true population proportion of 0.60. Use this data to estimate the probability that a confidence interval constructed in this way *will* include the true population proportion. Explain what it means to say that one is 95% confident of predictions made from polling data. Print your calculations and explanation.

Exercise 12.7 Repeat the steps in Exercise 12.4, but this time construct 90% confidence intervals for a population in which 60% of voters support the candidate (replace the multiplier 1.96 by 1.645).
(a) Follow the instructions in Exercise 12.5 to draw a chart of the confidence intervals.
(b) Repeat Exercise 12.6 to estimate the probability that a confidence interval constructed in this way *will* include the true population proportion.
(c) How do these intervals differ from those constructed in the previous exercise?
(d) Explain what it means to say that one is 90% confident of predictions made from polling data.
(e) What are the advantages and disadvantages of using a 90% confidence interval instead of 95% confidence interval in election polling.
(f) Print your calculations and discussions.

Exercise 12.8 Repeat the steps in Exercise 12.4, but this time construct 99% confidence intervals for a population in which 48% of voters support the candidate. In constructing the intervals replace the multiplier 1.96 by 2.576.
(a) Follow the instructions in Exercise 12.5 to draw a chart of the confidence intervals.

(b) Repeat Exercise 12.6 to estimate the probability that a confidence interval constructed in this way *will* include the true population proportion.

(c) In a text box explain what it means to say that one is 99% confident of predictions made from polling data.

(d) What are the advantages and disadvantages of using a 99% confidence interval instead of 95% confidence interval in election polling.

(e) Print your calculations and discussions.

Why Use $p = 0.5$ to Estimate Sample Standard Deviations?

Exercise 12.9 In this exercise you will construct a table of standard deviations for a variety of population proportions and sample sizes.

(a) Select a new worksheet and type in the table headings shown next.

	A	B	C	D	E	F	G	H	I	J	K	L
1		The Standard Deviation of Sample Proportions										
2					population proportion							
3			0.1	0.2	0.3	0.4	0.5	0.6	0.7	0.8	0.9	1
4		10										
5	sample	100										
6	size	1000										
7		1600										

(b) Type the formula =SQRT(C\$3*(1–C\$3)/\$B4)) into cell C4 as shown next.

			= =SQRT(C\$3*(1-C\$3)/\$B4)	
	A	B	C	D	E
1			The Standard Deviation of S	
2					p
3			0.1	0.2	0.3
4		10	0.095		
5	sample	100			
6	size	1000			
7		1600			

The value returned, 0.095, is the standard deviation of the sample proportions for samples of size 10 drawn from a population where the proportion is 0.1.

(c) Copy this formula to cell L4: Click in cell C4, move the cursor to the lower right corner of the cell, and when the cursor becomes a black cross, click and drag to cell L4. Explain what the entries in this row mean.

(d) Copy these formulas to the range C4:L7: Move the cursor to the lower right corner of Cell L4. When the cursor becomes a black cross, double-click. You should obtain the result shown next.

	A	B	C	D	E	F	G	H	I	J	K	L
1		The Standard Deviation of Sample Proportions										
2						population proportion						
3			0.1	0.2	0.3	0.4	0.5	0.6	0.7	0.8	0.9	1
4		10	0.095	0.126	0.145	0.155	0.158	0.155	0.145	0.126	0.095	0
5	sample	100	0.03	0.04	0.046	0.049	0.05	0.049	0.046	0.04	0.03	0
6	size	1000	0.009	0.013	0.014	0.015	0.016	0.015	0.014	0.013	0.009	0
7		1600	0.008	0.01	0.011	0.012	0.013	0.012	0.011	0.01	0.008	0

Exercise 12.10 Answer each of the following questions concerning the table constructed in Exercise 12.9.

(a) Explain why the $ signs were placed as they were in the formula entered in cell C4.

(b) Why are all of the entries in the right-most column of the table 0?

(c) We are told that the most conservative estimate p, the population proportion, for the purposes of calculating the standard deviation of the sample proportions is $p = 0.5$. Explain why the table provides evidence for this assertion.

(d) Print the table and the text box.

WHAT TO HAND IN: A carefully selected print region containing just the answers to the exercises. Do *not* print entire worksheets.

Simulation and Macros

13

In This Chapter...

- Simulating the birthday problem
- MATCH
- Recording and running a macro

If there are 23 people in a room, what is the probability that at least two have the same birthday? Surprisingly, it is greater than 1/2. In this chapter we will study this probability by designing a simulation in which 23 birthdays are generated at random and checked for matches. Running this simulation over and over again will provide data from which the probability can be estimated. A process like this in which the same commands are executed repeatedly can be expedited with a recorded macro.

13.1 RANDOM BIRTHDAYS

Selecting One Birthday at Random

The following procedure will generate one birthday at random and place it in cell A1.

1. Open a fresh workbook and select cell A1.
2. Set the **Calculation** mode to **Manual**. (See Section 8.1.)
3. Enter the command =RANDBETWEEN(1,365).

A "random birthday" will appear in A1 as shown next.

	=	=RANDBETWEEN(1,365)		
	A	B	C	D
1	192			
2				

Actually, the result will be an integer between 1 and 365, but this number becomes a birthday if you interpret 1 as January 1st and 365 as December 31 (ignoring leap years). In this way, for example, the integer 192 corresponds to July 11.

Selecting 23 Birthdays at Random

1. Set the **Calculation** mode to **Manual**.
2. Enter the command =RANDBETWEEN(1,365) in cell A1.
3. Split the screen horizontally so that cells A1 and A23 are both visible.
4. Select cell A1, hold down the SHIFT key and select cell A23.
5. Pull down the Edit menu and select Fill and then Down.
6. Recalculate the 23 commands: Hold down the SHIFT key and press the F9 key.

We'd like to know whether there are any duplicate birthdays in this list. In other words, is there an the integer that appears more than once? A quick way to find out is to arrange the birthdays in increasing order. Duplicate birthdays will then appear as consecutive identical values. Following the instructions in the next section to search the list for duplicates.

Finding Duplicate Birthdays

1. Convert the contents of range A1:A23, which is now a list of RANDBETWEEN commands, to a list of numbers.
   (a) If the range A1:A23 isn't already highlighted, select it. Click cell A1 hold down the SHIFT key and click cell A23.
   (b) Pull down the Edit menu and select Copy.
   (c) Pull down the Edit menu again and select Paste Special and then Values. Click the OK button.
2. With the range A1:A23 still selected, click the left-hand sort key on the Standard Toolbar. The result will look something like this:

	A	B
1	1	
2	17	
3	25	
4	66	
5	82	
6	82	
7	111	
8	130	
9	153	
10	173	
11	180	
12	181	
13	185	
14	199	
15	262	
16	279	
17	296	
18	308	
19	313	
20	327	
21	333	
22	339	
23	346	

Now you can scan through column A to look for two consecutive values that are identical. In the example above there is one pair of duplicate birthdays: the number 82, which appears in both rows 5 and 6, corresponds to a March 23rd birthday.

This particular list *did* produce a match. Was this just good luck or is such a duplication likely to happen again? In order to answer this question, we need to generate many, many lists of 23 birthdays, scanning each for duplicates. This could become a very tedious process and, worse yet, subject to human error. Eyes play tricks—it is easy to miss a pair or to see one that is not really there. So let's automate this scanning step using the fact that if two consecutive birthdays are identical, then their difference is 0. Fill column B with the differences between one column A entry and the next.

3. Select B1, enter the formula =(A2−A1).

4. Select the range B1:B22—click in cell B1, hold down the SHIFT key and click cell B22.

5. Pull down the Edit menu and select Fill and then Down.

6. Hold down the SHIFT key and press F9.

7. Since a zero in column B will indicate a duplicate birthday in column A, the next step is to look for a zero in column B. In C1 enter the following command =MATCH(0,B1:B23,0).

The result should look something like this:

	A	B	C	D
			=MATCH(0,B1:B22,0)	
1	1	16	5	
2	17	8		
3	25	41		
4	66	16		
5	82	0		
6	82	29		
7	111	19		
8	130	23		
9	153	20		
10	173	7		
11	180	1		
12	181	4		
13	185	14		
14	199	63		
15	262	17		
16	279	17		
17	296	12		
18	308	5		
19	313	14		
20	327	6		
21	333	6		
22	339	7		
23	346			

The number 5 in cell C1 indicates that the fifth element of the data range searched was a 0. This means that duplicate birthdays were found in cells B5 and B6.

The MATCH command searches a cell range for a particular value and if it finds the value, it returns the relative row or column where the match was discovered. If no match is found, Excel will print "#N/A." MATCH has three arguments. The first is the value it is looking for, in this case 0. The second is the range where it is to search, here B1:B23. The third argument indicates the type of match desired. Zero stands for an exact match. The number −1 or 1 in the third position means than an inexact match is acceptable.

13.2 MULTIPLE SIMULATIONS

In the preceding simulation a duplicate was found in 23 randomly chosen birthdays. This will not always be the case. Sometimes no match will be found, in other words, the 23 "persons" will have a different birthdays. The problem is to estimate the probability of at least one match. This can be done by repeating the simulation over and over again keeping track of the number of executions and of the number of times a duplicate birthday is found. The ratio of the number of duplicates to the total number of executions will approximate the probability of a duplicate birthday in a group of 23 people. To get a good estimate, the simulation should be carried out many times. This sort of thing is, to say the least, very tedious. There is, of course, a shortcut: You can run through the simulation procedure just once, recording your work as you go in a special Excel document called a *macro* and, from then on, a single keystroke will repeat the simulation.

WARNING!

Experienced cooks know that it is always a good idea to read through the recipe for a new dish a time or two before the actual cooking begins. The same is true of macro recording. Don't start typing until you have carefully studied the instructions for recording a macro in either Section 13.3 or Section 13.4, as appropriate for your version of Excel. Keep in mind that once you have clicked the OK button in the Record dialog box, every single keystroke and mouse click will be recorded and, further, that recording will not cease until a Stop Recording command is issued.

13.3 RECORDING A MACRO IN EXCEL 5 OR 7

Getting Ready

1. Open a fresh worksheet. Split the screen horizontally so that rows 1 and 23 are both visible.
2. Set the **Calculation** mode to **Manual**.
3. Pull down the Tools menu and select Record Macro and make sure that Use Relative References is checked. If it is not checked, select it.
4. Pull down the Tools menu again, select Record Macro and Record New Macro... .
5. When the Record New Macro dialog box appears, click on **Options>>**.
6. Type the macro name, say, "birthday" into the **Macro Name** box as shown below.

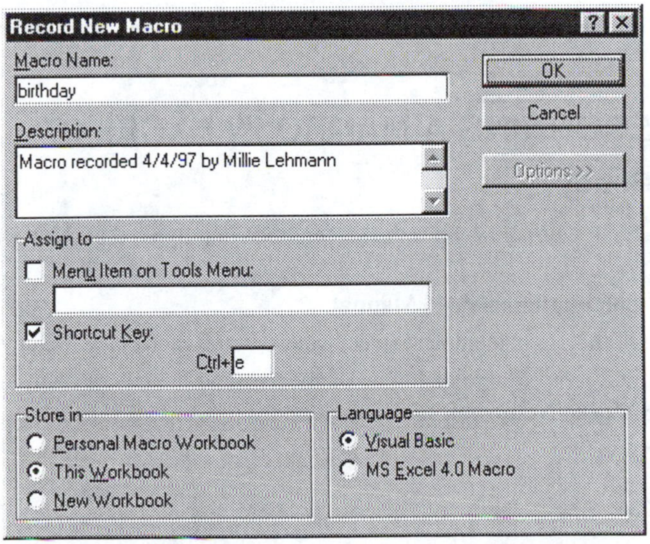

7. Under **Assign to** select **Shortcut Key**. Make a note of the shortcut key, probably **CTRL-e** as shown above. (On a Macintosh machine the typical keystroke sequence is **Option-Command-e**.) A letter other than "e" may be entered if you wish. This keystroke sequence is important because you will use it later to launch runs of the macro.

    Once you click the OK button, every command you type will be recorded. So, be careful. Do not click OK until you are ready to record the steps for generating 23 random birthdays.

8. Click the OK button.

Recording

1. Enter =RANDBETWEEN(1,365) into cell A1.
2. Select cell A1, hold down the SHIFT key and select cell A23.
3. Pull down the Edit menu and select Fill and then Down.
4. Hold down the SHIFT key and press the F9 key.
5. Pull down the Edit menu and select Copy.
6. Pull down the Edit menu again and select Paste Special and then Values. Click the OK button.
7. Click the left-hand sort key on the Standard toolbar.
8. Click cell B1 and enter the formula =(A2−A1).
9. Click in cell B1, hold down the SHIFT key and click cell B22.
10. Pull down the Edit menu and select Fill and then Down.
11. Hold down the SHIFT key and press F9.
12. In C1 enter the command =MATCH(0,B1:B23,0).
13. Pull down the Tools menu. Select Record Macro and then Stop Recording.

13.4 RECORDING A MACRO IN EXCEL 97

Getting Ready

1. Open a fresh worksheet. Split the screen horizontally so that rows 1 and 23 are both visible.
2. Set the **Calculation** mode to **Manual**.
3. Pull down the Tools menu and select Macro and Record New Macro... .
4. When the Record Macro dialog box appears, type the macro name, say, "birthday," into the **Macro Name** box as shown next.

5. In **Shortcut Key** type a letter, say "e." Make a note of the shortcut key, because the keystroke sequence **Ctrl-e** will launch the macro when it is complete.

   Once you click the OK button, every command you type will be recorded. So, be careful. Do not click OK until you are ready to record the steps for generating 23 random birthdays.

6. Click the OK button.

Recording

1. The Stop Recording Toolbar will appear on the screen. Make sure that the Relative Reference button on the Stop Rec toolbar is depressed as shown below.

2. Enter =RANDBETWEEN(1,365) into cell A1.
3. Select cell A1, hold down the SHIFT key and select cell A23.
4. Pull down the Edit menu and select Fill and then Down.
5. Hold down the SHIFT key and press the F9 key.
6. Pull down the Edit menu and select Copy.
7. Pull down the Edit menu again and select Paste Special and then Values. Click the OK button.
8. Click the left-hand sort key on the Standard Toolbar.
9. Click cell B1 and enter the formula =(A2−A1).
10. Click in cell B1, hold down the SHIFT key and click cell B22.
11. Pull down the Edit menu and select Fill and then Down.
12. Hold down the SHIFT key and press F9.
13. In C1 enter the command =MATCH(0,B1:B23,0).
14. Click the left button on the Stop Recording toolbar. Alternatively, pull down the Tools menu, select Macro and then Stop Recording.

13.5 TIPS

Using the Stop Recording Button

Macro recording can also be stopped by clicking the Stop Recording button that should appear on your screen as soon as the OK button is clicked in the Record New Macro dialog box. This button looks slightly different in the various versions of Excel. The Excel 5 button is shown on the left in the next illustration and the Excel 7 version on the right.

This button is actually an Excel toolbar, so if it doesn't appear automatically and you want to use it, you can select it from the View Toolbars menu before beginning the recording session.

If You Make a Mistake Recording a Macro

If you make a mistake before you've completed the steps of the macro, all is not lost.

1. Click the Stop Record Button or select Stop Recording from the Tools menu under Record Macro.
2. Select Macro... (or Macro) from the Tools menu.
3. When the Macro dialog box opens, click on the name of the stopped macro.
4. Click the Delete key.
5. Begin the recording process over again.

13.6 RUNNING THE MACRO

1. Open a new worksheet.
2. Click in cell A1.
3. Hold down the CTRL key and press the e key. (For the Macintosh, hold down the Option and the Command keys and press the e key.).
4. When the run is complete, click in cell D1 and run the macro again.

In order to determine whether there are duplicate birthdays in the simulation, just look at the top entry in the third column of output. If it's a number, there are duplicates. If it's "#N/A," there are no duplicates. The illustration below shows the result of four executions of the simulation. Two produced a match and two did not.

	A	B	C	D	E	F	G	H	I	J	K	L
1	15	12	#N/A	6	11	#N/A	26	3	2	2	5	4
2	27	10		17	6		29	0		7	20	
3	37	24		23	29		29	18		27	44	
4	61	30		52	11		47	8		71	0	
5	91	22		63	17		55	5		71	10	
6	113	3		80	18		60	12		81	0	
7	116	6		98	2		72	42		81	45	
8	122	30		100	45		114	16		126	5	
9	152	2		145	39		130	0		131	18	
10	154	29		184	9		130	6		149	42	
11	183	2		193	16		136	3		191	12	
12	185	2		209	11		139	10		203	4	
13	187	5		220	29		149	0		207	0	
14	192	11		249	23		149	39		207	5	
15	203	7		272	4		188	17		212	4	
16	210	34		276	15		205	43		216	14	
17	244	7		291	5		248	6		230	12	
18	251	6		296	21		254	21		242	6	
19	257	6		317	17		275	37		248	2	
20	263	15		334	3		312	26		250	25	
21	278	37		337	11		338	10		275	55	
22	315	6		348	5		348	9		330	11	
23	321			353			357			341		

EXERCISES

Exercise 13.1 Record the birthday macro. Print a worksheet like the one displayed, containing four executions of the macro.

Exercise 13.2 Now that you've got the macro set up, run 50 simulations of the birthdays of 23 people. What fraction of the simulations had duplicate birthdays? Print your calculations.

Exercise 13.3 Run 50 simulations of the birthdays of 10 people. What percentage of these simulations had duplicate birthdays? You can create another macro using the same method that you used to do the first problem. Print a copy of four executions of the macro together with the calculation of the percentage asked for.

WHAT TO HAND IN: Carefully select and print just the executions and calculations asked for. These will be large worksheets so avoid printing all of them.

Reporting *p*-Values

In This Chapter...

- Examples of hypothesis tests using the normal distribution
- Analysis Tools: *z*-test for the difference of means
- TDIST
- An Example of hypothesis testing using the *t*-distribution
- Analysis Tools: *t*-test for the difference of means

Hypothesis tests are a common technique for making statistical inferences from data. Such a test pits one claim about the distribution from which the data was sampled against another. One of these hypotheses says, essentially, that there is nothing noteworthy about this distribution. The second claims that the distribution has a special characteristic of some sort. The first of these is called the *null hypothesis*, usually symbolized H_0, and the second the *alternative hypothesis*, denoted H_a.

The test of the hypotheses always involves a probability calculation. It goes like this: Assume that the null hypothesis is true so that the population parameter in question has the value asserted by the null hypothesis. This assumption determines an expected value for the corresponding sample statistic. (For example, if we assume that the population mean is 0.5, then we expect the means of random samples to cluster near this value.) When the sample is collected and the data analyzed, the sample statistic will almost certainly deviate at least somewhat from its expected value. The question is—how extreme is this deviation? Could it be due to chance factors alone? This question is answered by calculating the probability of such a deviation being as extreme or even more extreme than that observed in the sample. This probability is called the *p-value* for the test. If the *p*-value is very small, then it is unlikely that the deviation observed in the sample is due to chance factors alone: It is much more likely that the null hypothesis is false. When this is the case, the null hypothesis is rejected in favor of the alternative hypothesis. A large *p*-value argues against acceptance of the alternative hypothesis. Excel provides many tools and commands for reporting *p*-values. We will look at some of them in this chapter.

14.1 USING THE NORMAL DISTRIBUTION TO TEST HYPOTHESES ABOUT SAMPLE MEANS

An Example of a One-Tailed Test for the Mean

Consider the following problem. A certain small private high school advertises that it does a superior job in preparing students for college entrance. How might this claim be statistically tested? One idea would be to study, say, the verbal SAT scores of graduating seniors to see if these scores are significantly higher on average than would be expected from national norms. SAT scores nationwide have a normal distribution with a mean of 500 and a standard deviation of 100. The distribution of these scores is displayed next.

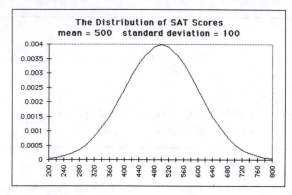

The question is: Are the SAT scores of the private school seniors, on average, significantly better than the nationwide average of 500? This question can be posed as a hypothesis test concerning the mean of the population from which these seniors are drawn.

H_0: The population mean is 500.

H_a: The population mean is > 500.

Open **datach14.xls** and select the sheet called **SAT scores**. It contains the SAT scores of the private school's senior class listed in the range B3:B102. (For the time being, ignore the data in column C. It will be looked at in the exercises.) Split the screen so that the first and last scores are visible in column B.

	A	B	C	D
1		verbal sat		
2		private	public	
3		538	432	
4		369	536	
97		486	458	
98		435		
99		486		
100		454		
101		495		
102		544		

1. **Give the data set a name.**
   (a) Select the data and its title: click in cell B2, hold down the SHIFT key and click in cell B102.
   (b) Pull down the Insert menu and select Name and then Create... .
   (c) In the Create Names dialog box check **Top Row** and then click the OK button.
2. **Calculate the sample average and size.**
   (a) Type labels for these statistics in cells A103:A104 as shown in the next illustration.
   (b) Enter the command =AVERAGE(private) into cell B103.
   (c) Enter =COUNT(private) into cell B104.

		=	=COUNT(private)	
		A	B	C
1			verbal sat	
2			private	public
3			538	432
4			369	536
97			486	458
98			435	
99			486	
100			454	
101			495	
102			544	
103	average		513.49	
104	n		100	
105				

3. **Calculate the standard deviation of the sample means.** Enter the command =100/SQRT(B104) into cell B105. The result of this calculation is shown next.

		=	=100/SQRT(B104)	
		A	B	C
1			verbal sat	
2			private	public
3			538	432
4			369	536
97			486	458
98			435	
99			486	
100			454	
101			495	
102			544	
103	average		513.49	
104	n		100	
105	sd samp means		10	
106				

Take stock of the situation before going on. If the Null Hypothesis is true, the 100 high school seniors can be viewed as a sample drawn from a population having

SAT scores with a mean of 500 and standard deviation of 100. The distribution of sample averages will, according to the calculations just made, have a mean of 500 and a standard deviation of 10. Given this assumption, we test the null hypothesis by answering the following question: What is the probability that such a sample mean will exceed the expected value of 500 by as much or more than this sample did? If this probability is small, we can comfortably reject the null hypothesis and accept the alternative hypothesis that the mean of the population from which the seniors were drawn exceeds 500. If the probability is large, then we will have no basis on which to reject the null hypothesis. The probability is pictured next. It is the area under the curve to the right of the vertical line.

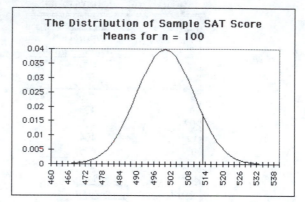

4. **Calculate the *z*-value.** The next step is to determine the standard or *z*-score associated with the sample mean value of 513.49. Enter the command =(B103–500)/B105 into cell B106. The result, shown next, is 1.349. This output counts the number of standard deviations that the sample mean lies above its expected value under the null hypothesis.

	A	B	C
		=(B103-500)/B105	
		verbal sat	
1			
2		private	public
3		538	432
4		369	536
97		486	458
98		435	
99		486	
100		454	
101		495	
102		544	
103	average	513.49	
104	n	100	
105	sd samp means	10	
106	z-value	1.349	
107			

5. **Calculate the *p*-value.** Calculate the probability of obtaining a sample mean value that exceeds its expected value by 1.349 standard deviations or more. This is the area under the standard normal curve from 1.349 to the right. The command =NORMSDIST(B106) will compute the area under the standard normal curve to the *left* of 1.349. Subtracting the result from 1 gives the area to the right. Enter the command =1−NORMSDIST(B106) into cell B107 as shown next.

	=	=1-NORMSDIST(B106)	
	A	B	C
1		verbal sat	
2		private	public
3		538	432
4		369	536
97		486	458
98		435	
99		486	
100		454	
101		495	
102		544	
103	average	513.49	
104	n	100	
105	sd samp means	10	
106	z-value	1.349	
107	p-value	0.0887	
108			

This output says that the probability is approximately 8.9% of obtaining a sample mean of 513.49 or higher when the expected value is 500. Is this probability "small enough" to permit rejection of the null hypothesis? The answer to the question is: It depends. Usually the rejection level is set before the study is conducted. Rejection levels of 1%, 5%, and 10% are common. In this case, if the researchers had set the level at 1% or 5%, they could not reject the null hypothesis, but if the level were set at 10% they could. The most useful course may be to simply report the *p*-value and let readers make their own interpretations.

An Example of a Two-Tailed Test

The test used in the previous section is an example of a *one-tailed test*, since the *p*-value was calculated from the right side of the distribution. We were only interested in whether the seniors had significantly *better* SAT scores than the national average.

In the quality control problem posed next, it is important to look at deviations from the expected value in both directions, so the test used will be *two-tailed*.

Open **datach14.xls** and select the sheet called **qual cont**. The range D2:D42 contains the diameters, in millimeters, of a sample of ball bearings taken from a large shipment. The sample average and size were calculated in cells D44 and D45, as shown next.

	=	=COUNT(Diameters)

	C	D	E
1		Diameters	
2		110.073	
3		110.040	
4		110.016	
38		110.026	
39		110.005	
40		109.905	
41		109.985	
42		109.977	
43			
44	average	110.011	
45	n	41.000	
46			

Suppose that the contract for the ball bearings specified a diameter of 110 mm for each ball bearing. This sample has an average diameter that differs from this value by 0.011 mm. Is this difference significant? Some deviation from this standard is to be expected due to both manufacturing and measuring inaccuracies, but the expected mean value is, nonetheless, 110 if the manufacturing process is properly tooled. Suppose that past production has shown a standard deviation in diameters of 0.038 mm, so that this number can be taken as the population standard deviation, in other words, the standard deviation in the diameters of all ball bearings produced. (When the population standard deviation is not known, the sample standard deviation is used in its place. See Exercise 14.5.) The distribution of sample means for samples of size 41 has a standard deviation of $0.038/\sqrt{41}$. The calculation is made in cell D46 as shown in the next illustration.

	=	=0.038/SQRT(D45)

	C	D	E
1		Diameters	
2		110.073	
3		110.040	
4		110.016	
38		110.026	
39		110.005	
40		109.905	
41		109.985	
42		109.977	
43			
44	average	110.011	
45	n	41.000	
46	sd sample means	0.0059346	
47			

These statistics can be used to test the relevant hypotheses.

H_0: The population mean is 110 mm.

H_a: The population mean is not 110 mm.

If the null hypothesis is true, then the distribution of sample means will be bell-shaped as shown next.

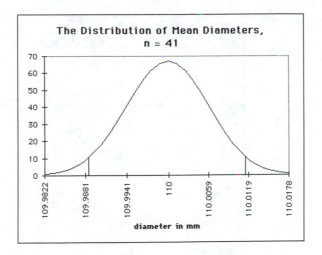

Vertical lines have been drawn at 110.011 mm and at 109.989 mm, both of which lie 0.011 mm from the expected mean value of 110 mm. The *p*-value for this hypothesis test is the probability that a sample mean will differ by 0.011 or more from the value, 110, specified by the null hypothesis. It is calculated by finding the total area in the two tails shown in the bell-shaped curve. Follow these steps to find this probability:

1. **Calculate the *z*-value.** Enter the command =(D44–110)/D46 in cell D47.

	C	D	E
		= =(D44-110)/D46	
1		Diameters	
2		110.073	
3		110.040	
4		110.016	
38		110.026	
39		110.005	
40		109.905	
41		109.985	
42		109.977	
43			
44	average	110.011	
45	n	41.000	
46	sd sample means	0.0059346	
47	z-value	1.892	
48			

2. **Calculate the *p*-value.** The command =NORMSDIST(D47) will compute the area under the standard normal curve to the left of 1.892. Subtracting this area from one gives the area to the right of 1.892. Doubling the result will compute the total area in the two tails. Enter the command =2*(1–NORMSDIST(D47)) into cell D48.

	C	D	E
		Diameters	
1			
2		110.073	
3		110.040	
4		110.016	
38		110.026	
39		110.005	
40		109.905	
41		109.985	
42		109.977	
43			
44	average	110.011	
45	n	41.000	
46	sd sample means	0.0059346	
47	z-value	1.892	
48	p-value	0.059	
49			

Formula bar: $= 2*(1-\text{NORMSDIST}(D47))$

The *p*-value is 0.059. This means that the probability of obtaining a sample average that differs from 110 by as much as this one did or more is 5.9% and, therefore, not small enough to reject the null hypothesis at the 5% level, although at a higher level it would justify a rejection of the null hypothesis. In real life such a decision is a judgement call.

EXERCISES

Exercise 14.1 Suppose that the senior class in the private high school described in the chapter had a different average verbal SAT score. For each of the average scores listed calculate the corresponding *p*-value.
(a) 510
(b) 525
(c) 517
(d) Discuss the *p*-values obtained in (a) through (c). In which does a rejection of the null hypothesis, as posed in the chapter, seem reasonable? Print your calculations and comments.

Exercise 14.2 Calculate the *p*-value for each of the following ball bearing samples. Use a standard deviation of 0.038.
(a) sample average = 110.015, sample size = 50.
(b) sample average = 109.997, sample size = 40.
(c) Discuss the *p*-values obtained above. In which does a rejection of the null hypothesis, as posed in the chapter, seem reasonable?
(d) Print your calculations and comments.

Exercise 14.3 The **SAT scores** worksheet in the workbook **datach14.xls** contains verbal SAT scores for a senior class from small public and private high schools located in the

same town.
(a) Suppose researchers wish to use the SAT scores to help determine whether the private school is doing a better job at college preparation. They use the following null hypothesis:

$$H_0 : \text{Population Mean}_{\text{private}} - \text{Population Mean}_{\text{public}} = 0$$

What is an appropriate alternative hypothesis?
(b) Open the worksheet and calculate the averages and sample sizes for both schools. Which school has the higher average SAT score? Do you think the difference is significant? Print a copy of your calculations and your thoughts about significance. In the next exercise, you will be asked to test the null hypothesis of no difference in the means against the alternative hypothesis that the mean SAT score for the private school population is the higher.

The Data Analysis Tool *z*-Test: Reporting *p*-Values for the Differences of Means

Exercise 14.4 This exercise builds on the previous one. If you have not already done it, do so now.
(a) Select Data Analysis... from the Tools menu. When the Data Analysis dialog box appears select *z*-**Test: Two Sample for Means** as shown next.

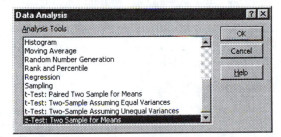

(b) Fill in the *z*-test dialog box as shown next.

The **Variable 1 Range** contains the private school SAT data and the **Variable 2 Range** the public school SAT data. The **Hypothesized Mean Difference** is the difference assumed by the null hypothesis, in this case **0**. **Labels** is checked because the variable ranges contain the labels "private" and "public" in row 2. **Variable 1 Variance** is 10,000, the square of the SAT standard deviation. **Variable 2 Variance** is also 10,000. **Alpha** refers to the maximum *p*-value that will be regarded as significant. The default value is 0.05, which will do for this problem. The **Output Range** specifies the upper left corner of the range where Excel's analysis of the data to be printed. When the dialog box has been filled in as illustrated, click the OK button and the following results will appear in the SAT score worksheet.

	A	B	C	D	E	F	G	H
1		verbal	sat					
2		private	public					
3		538	432					
4		369	536					
97		486	458					
98		435			z-Test: Two Sample for Means			
99		486						
100		454					*private*	*public*
101		495			Mean	513.49	512.25	
102		544			Known Variance	10000	10000	
103	average	513.49	512.25		Observations	100	95	
104	n	100	95		Hypothesized Mean Difference	0		
105					z	0.0864		
106					P(Z<=z) one-tail	0.4656		
107					z Critical one-tail	1.6449		
108					P(Z<=z) two-tail	0.2328		
109					z Critical two-tail	1.96		
110								

As you scan the results, you can see that Excel has made all of the relevant computations for you. The averages and counts found earlier have been recalculated and are found in rows 101 and 103 of column E. The variances are given in row 102 and the null hypothesis in row 104. The z-value for the observed difference of sample means is found in cell F105. The one-tailed p-value appears in cell F106[1] and is $p =$ 0.4656.

(c) Discuss the p-value. What does it mean and what does it say about the hypothesis test?

(d) Print a copy of Excel's z-test results along with your answers to the previous questions.

Reporting *p*-Values Using the *t*-Distribution

Exercise 14.5 When the standard deviation of the population is not known, the t-distribution rather than the normal distribution is used to calculate p-values. The mechanics in Excel are similar to those for the normal distribution except that the probabilities are calculated using the command TDIST and the standardized value used in the command is a t-statistic rather than a z-statistic.

In the ball-bearing example we assumed a known standard deviation of .038 mm. In real life such population parameters are seldom known. In this circumstance, the sample standard deviation replaces the unknown population standard deviation in standardizing the sample mean. Follow the steps below to see how Excel commands work for the t-distribution.

(a) Select the **qual cont** worksheet from the **datach14.xls** workbook. Copy the diameter data into a new worksheet and calculate the average and standard deviation of the diameters as well as the sample size so that the worksheet looks something like the one shown next.

	=	=STDEV(Diameters)	
	A	B	C
1		**Diameters**	
2		110.073	
3		110.040	
4		110.016	
40		109.905	
41		109.985	
42		109.977	
43			
44	average	110.011	
45	n	41.000	
46	samp sd dev	0.035	
47			

---

[1]Since the z-value is a positive number, the directions of the inequalities appearing in E106 and E108 must be reversed. Had the z-value been negative, the inequalities would be correct as they are written. (Another little Excel nuisance that must be borne with.)

(b) In cell D47 enter the command for the *t*-statistic for standardizing the difference of the sample mean from its expected value of 110, namely, (D44–110)/(D46/SQRT(D45)). The output is shown next.

		=	=(B44-110)/(B46/SQRT(B45))	
	A		B	C
1			**Diameters**	
2			110.073	
3			110.040	
4			110.016	
40			109.905	
41			109.985	
42			109.977	
43				
44	average		110.011	
45	n		41.000	
46	samp sd dev		0.035	
47	t-value		2.07187995	
48				

(c) The next step is to calculate the probability of obtaining a *t*-value greater than 2.07187995 or less than its negative. (Remember that the test must be two-tailed.) Probabilities associated with the *t*-distribution can be found using the command TDIST whose form as follows:

$$=\text{TDIST}(t\text{-value}, \textit{degrees of freedom}, \textit{number of tails})$$

Notice that it has three arguments. The first is the *t*-value, which in this case is located in cell B47. The second, degrees of freedom, is a parameter that determines the shape of the particular *t*-distribution to be used in calculating the probability. For tests of sample means, like this one, the number of degrees of freedom is the sample size minus 1, in this case, B45–1. The number of tails is, of course, 2.

(d) To calculate the *p*-value needed, enter the command =TDIST(B47,B45–1,2) into cell B48. The result is shown next.

	=	=TDIST(B47,B45-1,2)	

	A	B	C
1		**Diameters**	
2		110.073	
3		110.040	
4		110.016	
40		109.905	
41		109.985	
42		109.977	
43			
44	average	110.011	
45	n	41.000	
46	samp sd dev	0.035	
47	t-value	2.07187995	
48	p-value	0.045	
49			

(e) Open a text box and write a discussion of the meaning of this probability. Does it help resolve the hypothesis test? Print a copy of your calculations and your remarks about *p*.

Exercise 14.6 Data Analysis contains a tool called *t*-**Test: Two-Sample Assuming Unequal Variances**. This tool works very much like the *z*-test tool discussed earlier and should be used when the population standard deviations are not known. Use the tool for this exercise. Begin by selecting the worksheet entitled **HS GPA Coll** from the **datach14.xls** workbook.

(a) Is the average high school grade point average of arts majors significantly different from that of science majors? Pose this question as a hypothesis test and compute the appropriate *p*-value. Discuss your result. Print a copy of the *t*-test output and your discussion.

(b) Is the average high school grade point average of arts majors significantly different from that of non-arts majors? Pose this question as a hypothesis test and compute the appropriate *p*-value. Discuss your result. Print a copy of the *t*-test output and your discussion.

(c) Pose your own question concerning this data and construct an appropriate test. Print your calculations and conclusions.

Exercise 14.7 Use the tool **t-Test: Two-Sample Assuming Unequal Variances** in Data Analysis for this exercise. Open the worksheet entitled **pulse-gender-exercise** from the **datach14.xls** workbook. To answer the questions below, you will have to sort this data using either IF commands or AutoFilter.

(a) Is the pulse rate of men significantly different from that of women? Pose this question as a hypothesis test and compute the appropriate *p*-value. Discuss your result and print it.

(b) Is the pulse rate of exercisers significantly different from that of nonexercisers? Pose this question as a hypothesis test and compute the appropriate *p*-value. Discuss your result and print it.

(c) Pose your own question concerning this data and construct an appropriate test. Print your calculations and conclusions.

Exercise 14.8 The command RAND() selects a number between 0 and 1 at random. Use this command to generate 100 random numbers. These numbers can be viewed as a sample of 100 taken from a population of possible values that has a mean of 0.5 and a variance of 1/12. Is the average of your sample significantly different from its expected value of 0.5? Pose this question as a hypothesis test and compute the appropriate *p*-value. Discuss your result. Print your calculations and conclusions.

WHAT TO HAND IN: Print a minimum number of nicely organized pages with calculations labeled and discussions in text boxes. Select the print region carefully as there is no need to print the raw data.

Contingency Tables and Chi-Square

15

In This Chapter...

- Using Pivot Wizard to build a contingency table
- Constructing a table of expected values
- =CHITEST

The worksheet pictured below shows a contingency table of gender and smoking habits. It is displayed beginning in row 252. The table was constructed from the data listed in columns A and B using Pivot Wizard, Excel's built-in table-generating utility. The raw data was originally collected in a survey of 247 statistics students.

	A	B	C	D	E
1	male=1, female=2	Yes = 1, No = 2	right = 1, left = 2		
2	Gender	Smoker?	Handedness		
3	1	2	1		
4	1	2	2		
5	1	2	1		
246	1	2	1		
247	2	1	1		
248	2	2	1		
249	1	2	1		
250					
251					
252	Count of Smoker?	Smoker?			
253	Gender	Yes	No	Grand Total	
254	Male	26	64	90	
255	Female	40	117	157	
256	Grand Total	66	181	247	
257					

15.1 READING A CONTINGENCY TABLE

Take a few minutes to make sure that you understand the connection between the table in range A252:D256 and the data in range A3:B249. Each row in the data range gives the responses for one of the students surveyed, so, for example, the student who gave the row 4 responses was a left-handed male nonsmoker. The column headings, Gender, Smoker?, and Handedness, are called *fields*. Each of these fields has two *categories*—male or female, smoker or nonsmoker, and right or left handed.

Turning to the table (the handedness responses were ignored in compiling the table), consider the number 26 in cell B254. It is in the Male row and the Yes column of the contingency table. This entry means that in scanning the data in columns A and B, Excel found 26 entries that paired a 1 in column A with a 1 in column B; in other words, 26 male smokers. The number 157 in cell D255, to look at another example, is the grand total of all female respondents both smokers and nonsmokers. The body of the table has two rows, one for male and one for female and two columns, one for smoker and one for nonsmoker and is referred to as a *two-by-two* contingency table, denoted 2×2. A table with three row data categories and five column data categories would be denoted 3×5.

A contingency table is a wonderfully efficient way to present relationships between categorical data. Follow the steps in the next section to see how a contingency table is produced in Excel.

15.2 USING PIVOTTABLE WIZARD

The instructions in this section apply to Excel 5 and 7. Instructions for the version of Pivot Wizard contained in Excel 97 can be found in Appendix D.

PivotTable Wizard is Excel's table-building utility. It is usually referred to by the shortened title "Pivot Wizard."

Getting Ready

1. Open the workbook **datach15.xls** and click on the tab for the worksheet entitled **gender**. Copy the data to a new workbook and split the screen horizontally so that rows 1 and 249 are visible.

2. Highlight the range of data, A1:C249, as shown next. (Click in cell A3, hold down the SHIFT key, and click in cell C249.) You need not include the rows of headings, rows 1 and 2; Pivot Wizard automatically knows to include them. (Note: The data on handedness will not be used in this example, but it does not hurt to include it in the highlighted range and doing so will help you to better understand how Pivot Wizard works.)

	A	B	C	D
1	male=1, female=2	Yes = 1, No = 2	right = 1, left = 2	
2	Gender	Smoker?	Handedness	
3	1	2	1	
4	1	2	2	
5	1	2	1	
246	1	2	1	
247	2	1	1	
248	2	2	1	
249	1	2	1	
250				

Step 1 of Pivot Wizard

Pull down the Data menu and select PivotTable... . A dialog box containing the first step of Pivot Wizard will open. Make sure that **Microsoft Excel list or database** is selected as it is in the following picture. Click on the Next button.

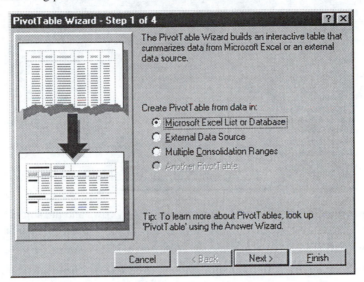

Step 2 of Pivot Wizard

1. Inspect the **Range** box. (It will look like the next illustration.) Notice that Excel has automatically included the row headings in the range. Correct the range entry if necessary.

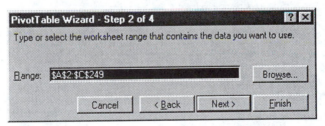

2. Click the Next button.

Step 3 of Pivot Wizard

You have reached the heart of Pivot Wizard. In this step, you actually design the table by telling Excel which fields to use for the rows and columns of the table and by specifying a format for the presentation of the body of the table.

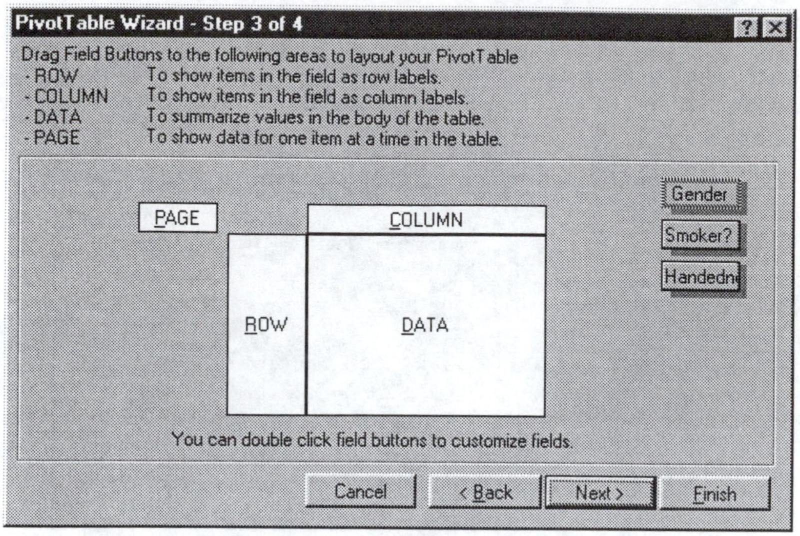

PivotTable Wizard - Step 3 of 4

Drag Field Buttons to the following areas to layout your PivotTable
- ROW   To show items in the field as row labels.
- COLUMN   To show items in the field as column labels.
- DATA   To summarize values in the body of the table.
- PAGE   To show data for one item at a time in the table.

PAGE   COLUMN   Gender
                                  Smoker?
ROW   DATA   Handedne

You can double click field buttons to customize fields.

Cancel   < Back   Next >   Finish

The table you are designing lists the gender categories by rows and the smoker categories by columns. The body of the table consists of *counts* of the number of students that fall into each of the four cells in the body of the table.

1. Click on the **Gender** button and drag it to the word **ROW**. The result should look like the next illustration.

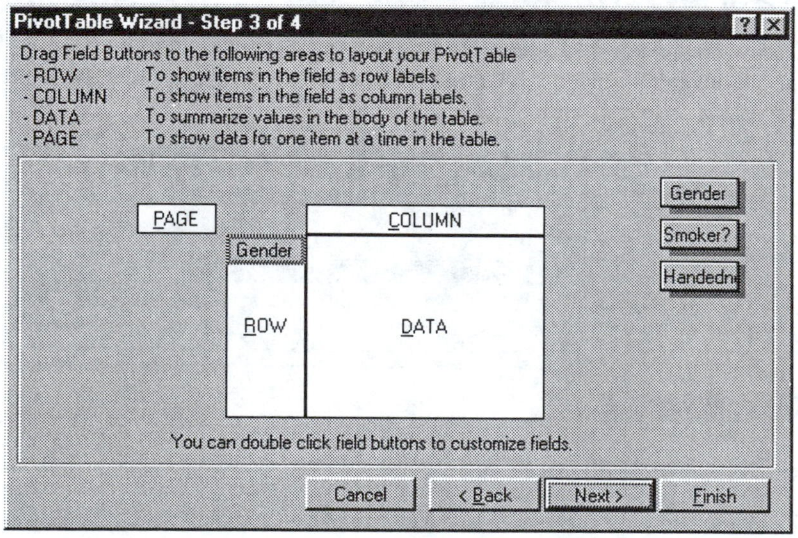

PivotTable Wizard - Step 3 of 4

Drag Field Buttons to the following areas to layout your PivotTable
- ROW   To show items in the field as row labels.
- COLUMN   To show items in the field as column labels.
- DATA   To summarize values in the body of the table.
- PAGE   To show data for one item at a time in the table.

PAGE   COLUMN   Gender
            Gender   Smoker?
                                  Handedne
ROW   DATA

You can double click field buttons to customize fields.

Cancel   < Back   Next >   Finish

2. Click on the **Smoker?** button and drag it to the word **COLUMN**. The result is shown next.

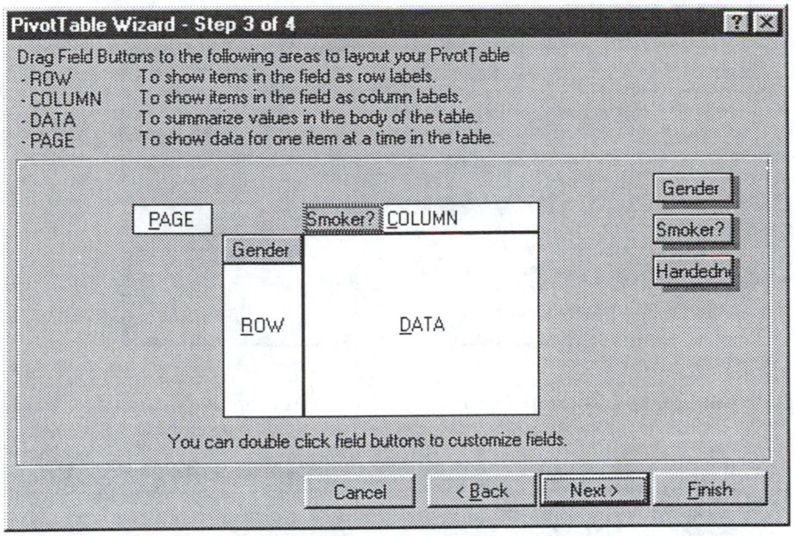

3. Click on either the **Smoker?** button or the **Gender** button and drag it to the word **DATA**. Excel will indicate the default presentation format for the data, **Sum of Smoker?**, as shown next.

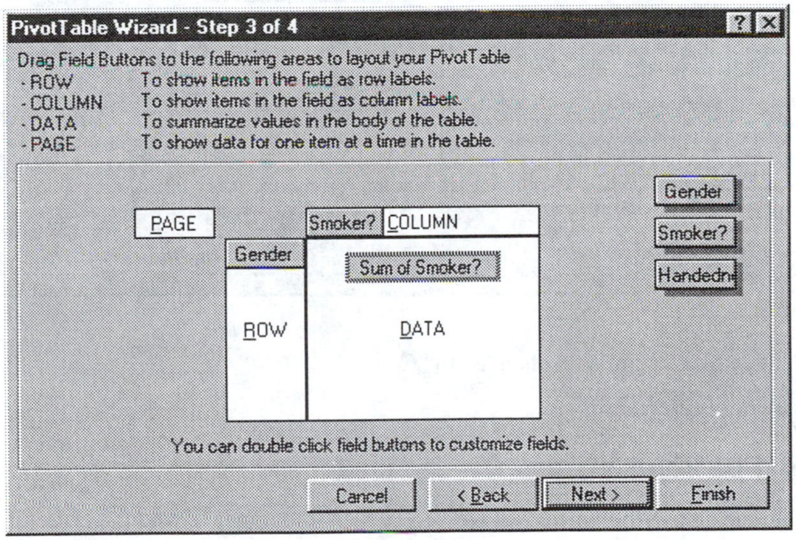

This is *not* what we want; **Sum** must be changed to **Count**.

4. Double-click on the word **Sum** to open the The PivotTable Field dialog box. When it appears click the word **Count** in the **Summarize by** window as shown next.

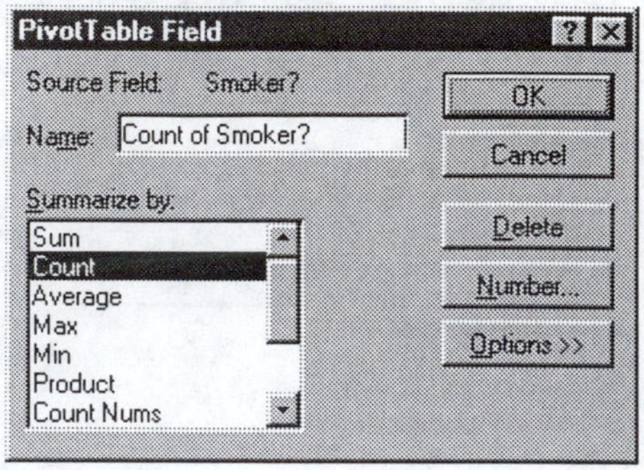

5. Click the OK button. Step 3 of Pivot Wizard will reappear.

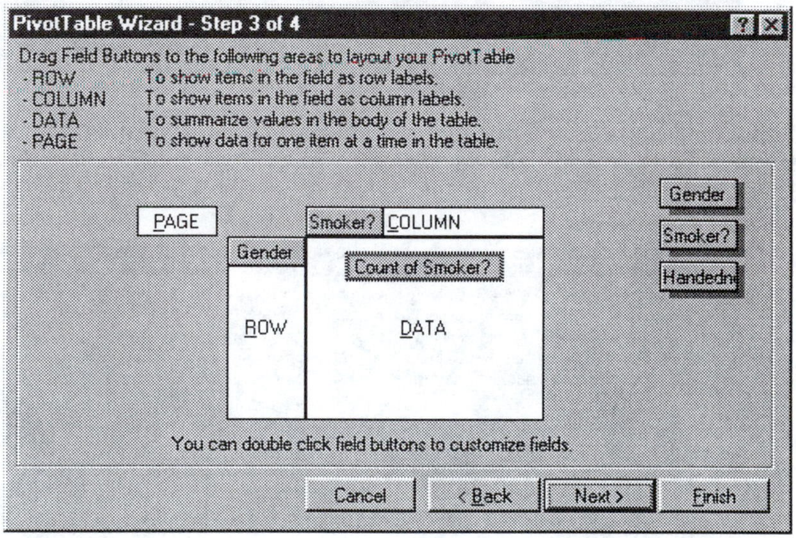

Notice that "Sum" was changed to "Count."

6. Click the Next button.

Step 4 of Pivot Wizard

Indicate the cell in the worksheet where you want the upper left-hand corner of the table located.

1. Click **Existing worksheet**.
2. Type in **A252** as shown below (or click the mouse on cell A252).

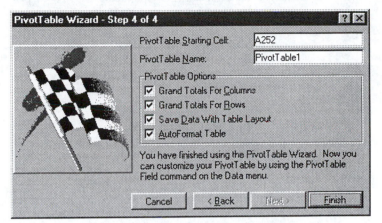

3. Click the Next button and Excel will embed the contingency table in the worksheet. A new toolbar, the Query and Pivot, may also appear. It should be ignored for the time being. If you want it out of your way, just click on its close box.

	A	B	C	D	E
1	nale=1, female=	= 1, No l = 1, left = 2			
2	Gender	Smoker?	landedness		
3	1	2	1		
4	1	2	2		
5	1	2	1		
246	1	2	1		
247	2	1	1		
248	2	2	1		
249	1	2	1		
250					
251					
252	Count of Smoker?	Smoker?			
253	Gender	1	2	Grand Total	
254	1	26	64	90	
255	2	40	117	157	
256	Grand Total	66	181	247	
257					

15.3 EDITING A TABLE PRODUCED BY PIVOT WIZARD

How to Alter Headings

Some entries in a Pivot Wizard table can be easily changed, but others cannot. Row and column headings are easy to modify. For example, the number 1 under Smoker? can be replaced with the word "Yes" by simply clicking in cell B253 and entering "Yes." The picture below shows the contingency table with renamed headings that make categories explicit.

	A	B	C	D	E
1	male=1, female=2	Yes = 1, No = 2	right = 1, left = 2		
2	Gender	Smoker?	Handedness		
3	1	2	1		
4	1	2	2		
5	1	2	1		
246	1	2	1		
247	2	1	1		
248	2	2	1		
249	1	2	1		
250					
251					
252	Count of Smoker?	Smoker?			
253	Gender	Yes	No	Grand Total	
254	Male	26	64	90	
255	Female	40	117	157	
256	Grand Total	66	181	247	
257					

Some Changes Cannot Be Made

It is *not* a simple matter to change entries in the body of the table because these counts are linked to the worksheet data entries. Excel will block an attempt to do so with an error message. For example, if you click on the number 26 in cell B254 and then press the DELETE key, Excel will refuse to erase the value and, instead, return an error message something like the one shown next.

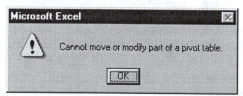

Click the OK button to recover.

When you attempt, for example, to delete one of the field names such as "Gender," you will also be scolded. The message box shown next is the sort of response Excel will give.

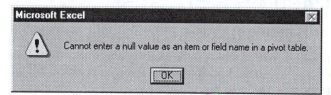

Click on the OK button.

Formatting Table Entries as Percentages

Some formatting can be altered. For example, the counts given in the body of the table can be changed to percentages of the total number of respondents as follows.

1. Click on one cell in the body of the table, such as B254.
2. Pull down the Data menu and select PivotTable Field... . Then click on **Options >>**. The PivotTable Field dialog box will open.
3. Click the down arrow for **Show Data As:** and select **% of total**. The dialog box should look something like the next illustration.

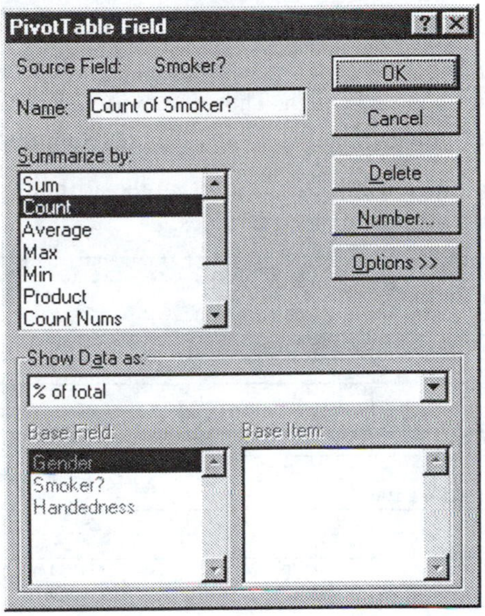

4. Click on the OK button, and the category counts will now be shown as percentages of the total number of respondents.

	Count of Smoker?	Smoker?		
252	Count of Smoker?	Smoker?		
253	Gender	Yes	No	Grand Total
254	Male	10.53%	25.91%	36.44%
255	Female	16.19%	47.37%	63.56%
256	Grand Total	26.72%	73.28%	100.00%

15.4 AN EXAMPLE OF A CHI-SQUARE TEST

In past decades smoking habits depended at least to some extent on gender: a greater proportion of men smoked than women. Does the contingency table computed earlier shed any light on this issue for the current generation of college students? (We will ignore, for the purposes of this example, the fact that these students do not constitute a random sample of all college students.) This section contains the steps for a Chi-Square test of the independence of gender and smoking habits.

Constructing a Table of Expected Frequencies

We begin by assuming that gender and smoking habits are independent and, based on this assumption, construct a 2×2 table of expected frequencies. We will then compare the actual frequencies to their expected values to see if a significant difference exists. If we find a difference that cannot be plausibly ascribed to chance, we will reject the assumption of independence.

As an example, calculate the expected number of male smokers. The fraction of men in the total sample is $90/247$. Based on the assumption of independence, we expect about the same fraction of smokers to be men. Thus, the expected number of male smokers is $66 \times (90/247)$, which rounds to 24. Recall that the actual number of male smokers is 26.

In the next picture, the range B261:C262 shows all the arithmetic calculations that must be made to complete the table of expected values.

	A	B	C	D	E
250	**Actual Values**				
251					
252	Count of Smoker?	Smoker?			
253	Gender	Yes	No	Grand Total	
254	Male	26	64	90	
255	Female	40	117	157	
256	Grand Total	66	181	247	
257					
258	**Expected Values**				
259		Smoker?			
260	Gender	Yes	No		
261	Male	=66*(90/247)	=181*(90/247)		
262	Female	=66*(157/247)	=181*(157/247)		
263					

In the next illustration, the numbers given in the calculations are replaced by their cell addresses in the Actual Values table.

	A	B	C	D	E
250	**Actual Values**				
251					
252	Count of Smoker?	Smoker?			
253	Gender	Yes	No	Grand Total	
254	Male	26	64	90	
255	Female	40	117	157	
256	Grand Total	66	181	247	
257					
258	**Expected Values**				
259		Smoker?			
260	Gender	Yes	No		
261	Male	=B256*(D254/D256)	=C256*(D254/D256)		
262	Female	=B256*(D255/D256)	=C256*(D255/D256)		
263					

The task before us is to design a command that can be typed into cell B261 and then simply copied to the range B261:C262. The idea is to avoid typing in each of these formulas individually. A clever use of absolute and relative cell addressing will do the trick. For example, the address of the divisor, D256, does not change from one cell to the next. This means that the formula must use an absolute address for this cell, namely, $D$256. Notice that two other references do not change: the row address 256 and the column address D. Therefore, the formula you need to enter in cell B261 is

$$=B\$256*(\$D254/\$D\$256)$$

The result is shown next.

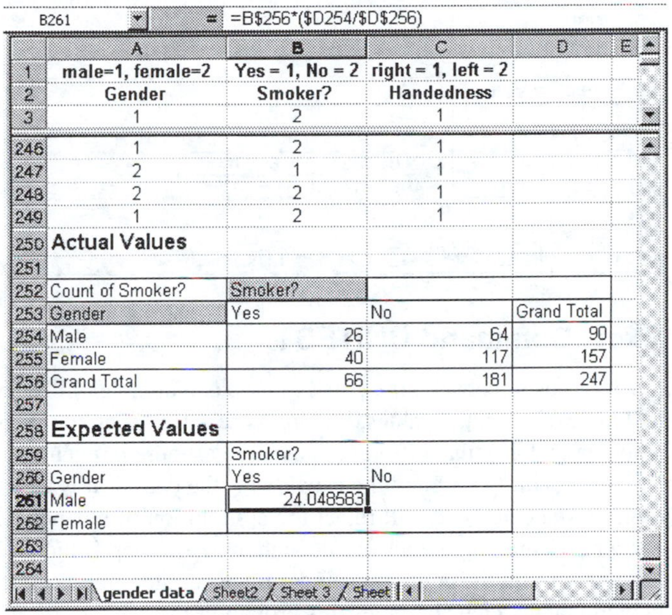

B261		= =B$256*($D254/$D$256)			
	A	**B**	**C**	**D**	**E**
1	male=1, female=2	Yes = 1, No = 2	right = 1, left = 2		
2	Gender	Smoker?	Handedness		
3	1	2	1		
246	1	2	1		
247	2	1	1		
248	2	2	1		
249	1	2	1		
250	**Actual Values**				
251					
252	Count of Smoker?	Smoker?			
253	Gender	Yes	No	Grand Total	
254	Male	26	64	90	
255	Female	40	117	157	
256	Grand Total	66	181	247	
257					
258	**Expected Values**				
259		Smoker?			
260	Gender	Yes	No		
261	Male	24.048583			
262	Female				
263					
264					

gender data / Sheet2 / Sheet 3 / Sheet

Use a black cross drag to copy this formula to the range B261:C262.

C262	▼	=	=C$256*($D255/$D$256)		

	A	B	C	D	E
1	male=1, female=2	Yes = 1, No = 2	right = 1, left = 2		
2	Gender	Smoker?	Handedness		
3	1	2	1		
246	1	2	1		
247	2	1	1		
248	2	2	1		
249	1	2	1		
250	**Actual Values**				
251					
252	Count of Smoker?	Smoker?			
253	Gender	Yes	No	Grand Total	
254	Male	26	64	90	
255	Female	40	117	157	
256	Grand Total	66	181	247	
257					
258	**Expected Values**				
259		Smoker?			
260	Gender	Yes	No		
261	Male	24.048583	65.951417		
262	Female	41.951417	115.048583		
263					
264					

Ⅰ◀ ▶ ▶Ⅰ \ **gender data** ∕ Sheet2 ∕ Sheet 3 ∕ Sheet ◀

Using the Excel Command CHITEST

There are clear differences in between the entries in the actual table of observed values
and the calculated table of expected values, but are these differences significant? We can
ask Excel to find the probability that differences as large as these or even larger are due to
chance factors. This probability is called the *p-value* for the data. If it is very small—less
than 5% is a common standard—we will reject the assumption of independence and
conclude that gender and smoking habits are dependent. However, if the probability
is larger than 5%, again using the common standard, we can not draw a conclusion of
dependency. The command which computes this probability is

=CHITEST(*actual range,expected range*)

When this test is applied to the gender and smoking data the *p-value* is 56%. The
calculation is shown next.

B265	▼	=	=CHITEST(B254:C255,B261:C262)		
	A	**B**	**C**	**D**	**E**
250	**Actual Values**				
251					
252	Count of Smoker?	Smoker?			
253	Gender	Yes	No	Grand Total	
254	Male	26	64	90	
255	Female	40	117	157	
256	Grand Total	66	181	247	
257					
258	**Expected Values**				
259		Smoker?			
260	Gender	Yes	No		
261	Male	24.048583	65.951417		
262	Female	41.951417	115.048583		
263					
264					
265	p-value	0.559852973			
266					
267					
268					
269					
270					
271					

gender data / all / gender-college / layoff data / E

So, it can't be concluded from this data that gender and smoking habits are dependent on one another in the population from which the sample was drawn. This suggests that the decision to smoke may not be as much influenced by gender as it once was.

EXERCISES

Data for these exercises are found in **datach15.xls** .

Exercise 15.1 Reproduce the test of the independence of smoking habits and gender outlined in Section 15.4. Print a copy of the actual and expected value tables and of the calculation of the p-value.

Exercise 15.2 The following questions refer to the **gender** worksheet.
(a) Construct a contingency table of gender and handedness.
(b) Compute a table of expected values.
(c) Does this data refute the assumption that gender and handedness are dependent? Use a Chi-Square test and answer the question in a text box.
(d) When you are finished with the Chi-Square test, convert the actual values table to percentages of the total.
(e) Print a copy of the tables, the calculation of the p-value, and the text box.

Exercise 15.3 Open the **gender-college** worksheet.
(a) Construct a 2×5 contingency table of gender and college.
(b) Compute a table of expected values.

(c) Does this data refute the assumption that gender and college choice are independent? Use a Chi-Square test.
(d) Copy the contingency table to a new worksheet using Paste Special...Values from the Edit menu. Delete the Business College data, correct the totals, and test this table for independence.
(e) Open a text box and write a paragraph discussing the two probabilities obtained.
(f) Print a copy of the tables, the calculation of the *p*-value, and the text box.

Exercise 15.4 Open the **layoff** data worksheet.
(a) Test the independence of race and layoff status using CHITEST.
(b) Open a text box and write a discussion of the results you obtained.
(c) Print a copy of the tables, the calculation of the *p*-value, and the text box.

Exercise 15.5 Open the **earnings+** worksheet.
(a) Make a 6 × 2 contingency table of earning level and gender.
(b) Calculate the table of expected values.
(c) Use CHITEST to decide whether there is evidence that these two variables are not independent.
(d) Print a copy of the tables, the calculation of the *p*-value, and a text box discussion of the test.

Exercise 15.6 Open the **earnings+** worksheet.
(a) Make a 5 × 6 contingency table of earning level and college.
(b) Calculate the table of expected values.
(c) Use CHITEST to decide whether there is evidence that these two variables are not independent.
(d) Print a copy of the tables, the calculation of the *p*-value, and a text box discussion of the test.

Exercise 15.7 Open the **earnings+** worksheet.
(a) Make a 2 × 5 contingency table of smoking habits and college.
(b) Calculate the table of expected values.
(c) Use CHITEST to decide whether there is evidence that these two variables are not independent.
(d) Print a copy of the tables, the calculation of the *p*-value, and a text box discussion of the test.

WHAT TO HAND IN: For each exercise assigned, print a worksheet selection that includes the contingency table, the table of expected values, and the CHITEST result plus any commentary asked for written in a text box. Do not print the data.

Answers to Selected Exercises

Chapter 1

1.2(b)   For example, in 1995 the female enrollment was 131% of the male enrollment.

1.2(c)   In evaluating a charge of gender discrimination it would be important to look at, among other factors, student recruitment practices and the gender breakdown in applicants.

1.3(a)   2046

1.3(b)   The formula is =B2/B$7, which when copied down gives, for example, a business enrollment of 24% of total enrollment.

1.4(e)   More than 10 times as many men, age 18–21, as women were arrested for alcohol related offenses. Clearly, this is a much more severe problem for young men.

1.5(d)

	A	B	C	D	E	F	G
1							
2	Alcohol Related Arrests by Age and Sex in Oklahoma, September to December 1973						
3		MALES			FEMALES		
4		18-21	Over 21	Total	18-21	Over 21	Total
5	Driving under influence	427	4973	5400	24	475	499
6	Drunkenness	966	13,747	14,713	102	1176	1278
7	Total	1393	18720	20113	126	1651	1777
8							
9	Percent of Alcohol Related Arrests by Age and Sex in Oklahoma, September to December 1973						
10		MALES			FEMALES		
11		18-21	Over 21	Total	18-21	Over 21	Total
12	Driving under influence	2%	25%	27%	1%	27%	28%
13	Drunkenness	5%	68%	73%	6%	66%	72%
14	Total	7%	93%	100%	7%	93%	100%

1.6   The 18–21 age group accounts for 7% of the total arrests for alcohol-related crimes and this proportion applies equally to men and women; hence it is unfair to permit women of this age to drink while forbidding it to men. As it turned out, the Supreme Court overturned the law. They were impressed by data that suggested that fewer young women drove, but that among all drivers the proportion of women driving under the influence of alcohol was only slightly lower than that of men.

Chapter 2

2.1(b)   mean = 3.05125, median = 3.05, mode = 3.00

2.2(a)   count is 272, 128 better, 122 average, 22 worse than average, mean rating is 1.6

2.3(a)   mean = 1.6, median = 2, mode = 1

2.3(b)   The median does not work well for this data since it suggests a set of balanced responses. The fact that most of the ratings were average and above is better indicated by either the mean or the mode.

2.4

	men	women
count	99	172
better	54	73
average	39	83
worse	6	16
mean rating	1.5	1.7

2.5(a)   mean = 3.46, median = 3.2, mode = 3

2.6(a)   mean = 1,878,476 miles, median = 2000 miles, mode = 1500 miles

2.6(b)   mean = 15,960 miles, median = 3000 miles, mode = 3000 miles

2.6(d)   Outliers created by wildly high guesses moved the average away from the median and the mode. In this case, the median is the preferred measure of central tendency.

2.7(a)   70%

2.7(b)   77%

2.8(a)   min = 0, first quartile = 3, median = 3.2, third quartile = 3.5, max = 34

2.8(b)   90th percentile = 3.8, 10th percentile = 2.53.

2.9(b)   66 for exercisers and 72 for nonexercisers

2.9(c)   for exercisers 52, 60, 64, 68, 88 and for nonexercisers 56, 68, 72, 80, 88

2.9(e)   average for men 68 and for women 74

2.9(f)   If the pulse rate for women is naturally higher than that for men, then the average pulse rate of any group can be raised or lowered according to the proportion of women in it.

Chapter 3

3.3(a) The test data contains two outlier values, −20 and 40, otherwise the data is roughly symmetric centered at 3.

3.3(b) The symmetric character of the data between 1 and 6 is not revealed by the default histogram.

3.6(a) $35,000 to $56,000

3.6(d) $50,000

3.6(c) $67,000 to $80,000

3.6(d) $40,000 to $45,000

3.7(a)

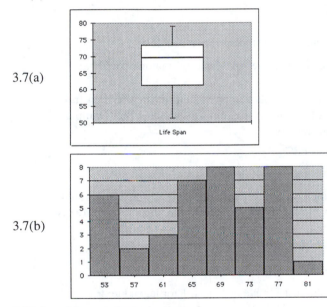

3.7(b)

3.8(a) Most students don't smoke.

3.9(a) A start value of 30, bin width of 10, and finish value of 80 produces a nice histogram, but other choices for these parameters can be justified.

3.10(b) Try setting the extreme percentile value at 2%.

3.11 Try setting the extreme percentile value at 2%.

3.12(a) A start value of 50, bin width of 2 and end value of 84 produces a nice histogram, but other choices for these parameters can be justified.

Chapter 4

4.3(a)   Jack McDowell

4.3(b)   John Dopson, for example

4.3(c)   22

4.3(d)   The lowest salary is the most frequently occurring salary.

4.5   Skewed to the left

Chapter 5

5.4(a)   Categories overlap because some students have double majors.

5.4(b)   To produce an accurate pie chart the "overlapping" majors would have to be known.

5.5(b)   In the early years there was a sharp improvement in speeds, then from 1956 to 1968 the times held steady. Beginning with the 1972 Olympics, speeds again improved steadily.

5.7(f)   Texas 32%, Other States 68%

5.7(g)   For example, 1982 50% and 1993 47%

5.7(h)   At the very least, the incidence of capital crime in Texas as compared to other states would have to be known, as well as the population of Texas compared to other states.

5.8(a)   Data series in columns, column 1 for x-axis labels, 0 for legend text

5.8(d)   Data series in rows, column 1 for x-axis labels, 1 for legend text

5.9(a)   Data series in rows, column 1 for x-axis labels, 1 for legend text

5.9(b)   Use column chart, Option 3

5.9(d)   Begin by calculating the daily sums. The pie chart should show Monday workouts, for example, at 17%.

Chapter 6

6.3(c)   For example, the (b) chart tends to exaggerate the gains made by women since the bar for 1990 is several times longer than the bar for 1950, whereas the percentage for 1990 is not even twice as large as that for 1950.

6.4(c)   For example, the (b) chart does a better job of showing the nearly parallel improvement in the times for men and women (perhaps due to improved training practices used by both).

6.5   Generally speaking, as the population increased so did the number of crimes. This means that the increase in the number of crimes could be at least partially due to the increase in population size.

6.6(a)   Enter the command =B4/B3 into cell B5 and then use a black cross drag to copy the formula across row 5.

6.6(b)   Enter the command =B5*100000 into cell B6 and then use a black cross drag to copy the formula across row 6.

6.6(c)   25%

Chapter 7

7.2(b) $R = 0.938$

7.2(c) $y = 0.9104x + 1.6195$, $R^2 = 0.8803$

7.2(d)   28.9 years

7.2(e)   There is a strong positive association between the variables suggesting that as the age of a bride increases so does the age of her groom.

7.3(b) $R = 0.359$

7.3(c)   1592 mm

7.4(b) $R = 0.007$

7.5(a) $y = 0.0036x + 0.5386$, $R^2 = 0.235$

7.6(a) $y = -19.209x + 337.4$, $R^2 = 0.3986$

7.6(b) $R = -0.0631326$

7.6(c)   There is a moderate negative association between the variables, indicating that as the number of hours of sleep required per day for mammals increases the gestation period decreases.

7.7 $y = 0.7687x + 11.161$, $R^2 = 0.9576$

7.8(e)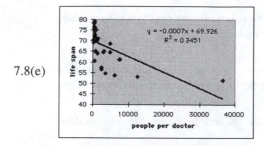

7.9(a)–(d) $y = 0.5x + 3, R^2 = 0.666$

7.9(e) (c)

7.10(b) $R = 0$

7.10(d) $R = 1$

7.11 The exponential trendline fits best, as $R = 1$.

7.12(b) The R^2 value changes from 0.54 to 0.89 with the removal of the USA point.

7.13(b) Man and the two elephants appear from the plot to have outlier status. The effect of their points on the plot should be studied.

Chapter 8

8.2(a) ± 1

8.2(b) Answers will vary, but in one case the percentages were 0% and 80%.

8.2(c) Answers will vary, but in one case the percentages were 20% and 15%.

8.2(d) Answers will vary, but in one case the percentages were 90% and 0%.

8.3(b) Perhaps 10, certainly 100.

8.5(b) Men with birth dates later in the year tended to have somewhat lower draft numbers, making them more liable to be drafted.

Chapter 9

9.1(a) 0.841345

9.1(b) 0.818595

9.1(c) 0.066807

9.1(d) 552.4401

9.2(a)

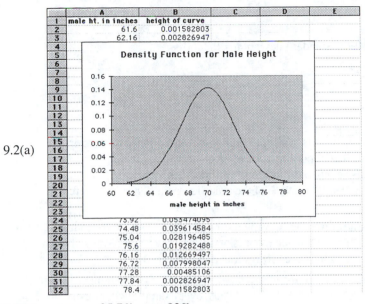

	A	B	C	D	E
1	male ht. in inches	height of curve			
2	61.6	0.001582803			
3	62.16	0.002826947			
4					
5					
6					
7					
8					
9					
10					
11					
12					
13					
14					
15					
16					
17					
18					
19					
20					
21					
22					
23					
24	73.92	0.053474095			
25	74.48	0.039614584			
26	75.04	0.028196485			
27	75.6	0.019282488			
28	76.16	0.012669497			
29	76.72	0.007998047			
30	77.28	0.00485106			
31	77.84	0.002826947			
32	78.4	0.001582803			

9.2(b)   women 25.7%, men 92%

9.3   Score must be 594 or higher.

9.4(b)   68.3%

9.4(c) $z = 0.6745$

9.4(d) $z = -0.84162$

9.6(a)   average = 64, standard deviation = 13.5

9.6(b)   4% is the normal curve approximation.

9.6(c)   3.6% is the actual value.

9.7(b)   average = 2.68, standard deviation = 3.32

9.7(c)   58% is the normal curve approximation.

9.7(d)   46% is the actual value.

9.8(g)   16% of the ages are below 52 years.

9.9

	The Empirical Rule			
	billionaire's wealth			
average minus 3sd	-7.2749759	0.15 th	percentile	1
average minus 2sd	-3.9561356	2.5 th	percentile	1
average minus 1sd	-0.6372953	16 th	percentile	1.2
average	2.68154506	50 th	percentile	1.8
average plus 1sd	6.00038538	84 th	percentile	3.8
average plus 2sd	9.3192257	97.5 th	percentile	10.34
average plus 3sd	12.638066	99.85 th	percentile	32.476

Chapter 10

10.6(a)   average = 0.4966, standard deviation = 0.2860

Chapter 12

12.6   When a pollster reports a sample proportion and a margin of error, the pollster is, in effect, reporting a confidence interval. In the long run, confidence intervals set up in the fashion described in this exercise will contain the population proportion 95% of the time. The confidence level refers to this probability.

12.10(a) The population proportions are always found in row 3 so the row address for p must be absolute, in other words, written with a $ sign. Similarly for the sample sizes, which are always found in column B.

12.10(b) Think of this in terms of the election poll example. If $p = 1$, it means that every voter intends to vote for Cole. Thus, every sample will consist entirely of Cole supporters and $\hat{p}$ will be 1 in every case, and there will be zero deviation of the sample proportions from the population proportion.

12.10(c) For all of the sample sizes represented in the table, the largest standard deviation for the sample proportions occurs when $p = 0.5$. Using $p = 0.5$ will give the worst-case value for the standard deviation.

Chapter 14

14.1(a) $p = 0.16$

14.1(b) $p = 0.006$

14.1(c) $p = 0.045$

14.2(a) $p = 0.005$

14.2(b) $p = 0.62$

14.3(a)   An appropriate alternative hypothesis would be that the private school population mean SAT score exceeds that for the public school.

14.3(b)   The sample mean for the private school is 513.49 whereas that for the public is 512.25.

14.6(a)   The one-sided p-value is 0.011.

Chapter 15

15.2(c) $p = 0.039$

15.3(c) $p = 0.041$

15.3(d) $p = 0.474$

15.4 $p = 3.3344 \times 10^{-16}$

15.5(c) $p = 0.078$

15.6(c) $p = 0.197$

15.7(c) $p = 0.243$

The Add-Ins

B

In This Appendix...

- Loading **STAT**
- Placing the **STAT** Add-Ins in Excel's Library
- Loading the Tools

If you are an experienced computer user and have moved and copied files from one folder or directory to another in the past, you probably will have no trouble with these instructions. However, if you are a real beginner, give these instructions a try, but be prepared to ask for help if you get frustrated and confused. What must be done is actually very straightforward. The first task is to install the compressed contents of **STAT**, the diskette that comes with this book, on the computer's hard drive (drive C on a Windows machine). This installation will place a directory or file folder called **STAT** (or, perhaps, **Stat**) on the computer's hard drive. The directory will contain all of the **datach** workbooks as well as a sub-directory called **seeaddin** containing three Excel add-ins. These add-ins must then be copied to Excel's Library file. That's all there is to it. Detailed instructions follow for a Windows 95 or 97 machine. If you are a Macintosh user, see B.4. If you are using this book as a lab manual for a college or university course, your instructor may have taken care of these installations for you. In this case, ask your instructor for instructions on retrieving the **datach** files.

B.1 HOW TO LOAD THE STAT DISKETTE ON A WINDOWS 95 OR 97 MACHINE

1. Place the **STAT** diskette in the floppy disk drive, probably drive A.
2. In Windows 95 or 97 select Run from the Start menu.
3. When the Run dialog box appears, type A:INSTALL (if your floppy drive has a different letter, type it in place of A).

4. The installation program will place the decompressed contents of **STAT** in a directory or file called **STAT** (or **Stat**) located on your computer's internal hard drive. While the program is working, an installation window will be open on the screen. When the installation is complete, the title of the window will read "Finished-INSTALL." Close the window by clicking on the "×" in its upper right corner.

B.2 PLACING THE ADD-INS IN EXCEL'S LIBRARY

Open the Add-Ins Window from STAT

1. Select Find from the Start menu and then select Files and Folders.

2. When the Find window opens, click the **Name and Location** tab.

3. Type **STAT** into **Named** and make sure that the **Look in** window shows **(C:)** (unless your hard drive has some other name) as shown next.

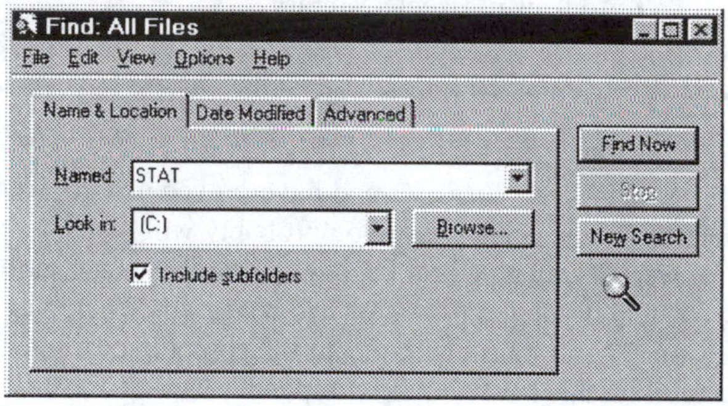

4. Press the **Find Now** button.

5. A list of files will appear at the bottom of the dialog box in the **Name** window. Look for a file folder called **Stat**.

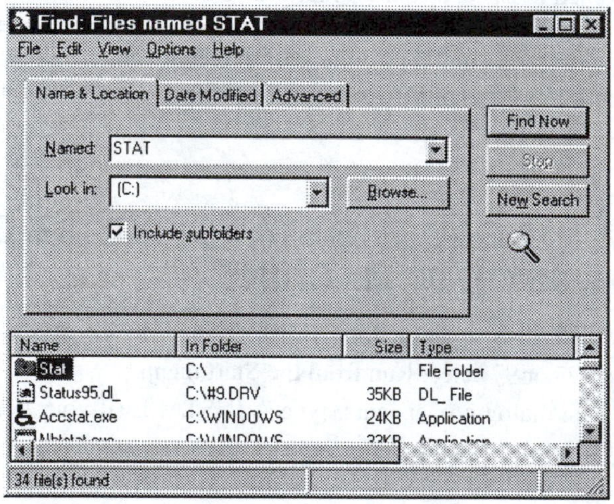

6. Double-click on the **Stat** folder.
7. Double-click on the **seeaddin** folder.
8. Leave the Find window open and leave the seeaddin window open.
9. Go on to the next section.

Open the Excel Library

1. Return to the Find dialog box.
2. Click in the **Named** box, delete STAT and type **Library**. Press the **Find Now** button.
3. A list of files will appear at the bottom of the Find dialog box. Look for a file folder called Library.

Its path name under **In Folder** will be something like C:\Program Files\Excel\Office or C:\MSOffice\Excel.

4. When you have located the Excel Library file folder, double-click on it.
5. Close the Find window.
6. Now you will have two windows open on the screen, the Add-In window and the Library window. Leave them both open.

Copy the Files from Add-In to Library

1. Select all of the files in the seeaddin window: Click on the first file in the list, hold down the SHIFT key, and click on the last file in the list.

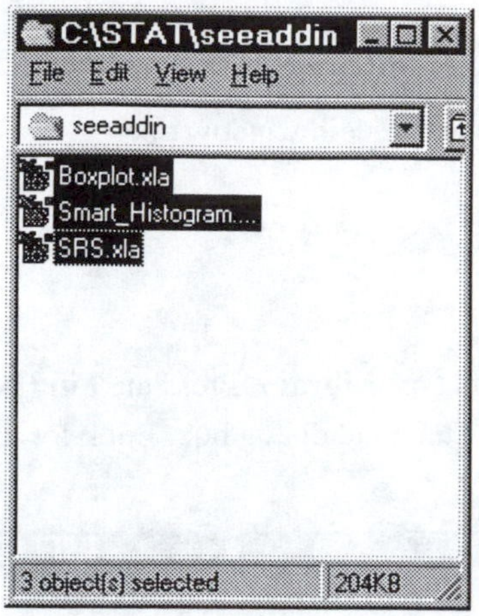

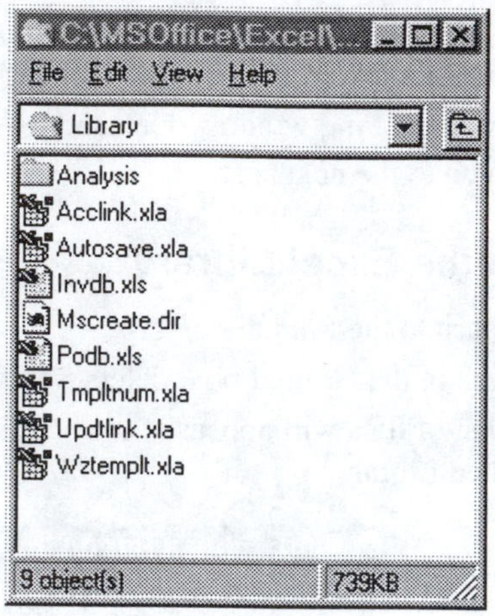

2. Hold down the CTRL key and drag the files to the Library window. The Library window should now contain the three add-ins as shown next.

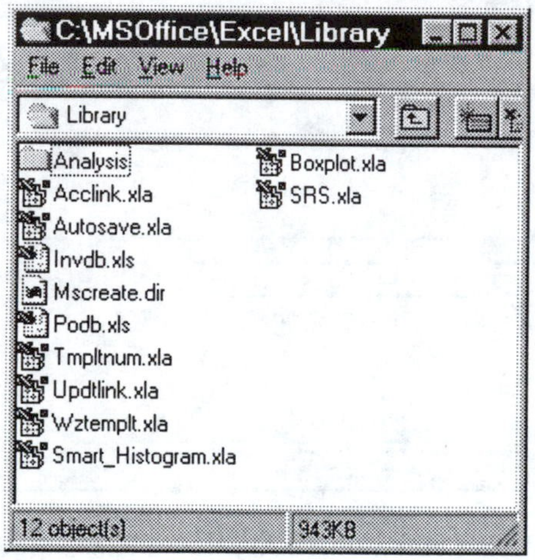

3. Close the windows.

B.3 LOADING THE STATISTICAL TOOLS

1. If you have not already done so, follow the instructions in Sections B.1 and B.2.
2. Open an Excel workbook, pull down the Tools menu and release on Add-Ins... .

3. When the Add-Ins dialog box appears, make sure that each of the following add-ins are checked:

    (a) Analysis ToolPak

    (b) Boxplot

    (c) Smart Histogram

    (d) Srs

4. Click the OK button.

What to Do If Some of the Add-Ins are Missing

If Analysis ToolPak is missing from the Add-In Available list, it means that the file was not loaded when Excel was installed on your machine. In this case, you will have to install it from the original Excel diskettes.

If Boxplot, Smart Histogram, or Srs are missing, follow the instructions in Sections B.1 and B.2 to load them.

B.4 ADD-INS FOR THE MACINTOSH

To obtain the STAT diskette formatted for the Macintosh contact Duxbury Press at the following address.

> Duxbury Press
> 10 Davis Dr.
> Belmont, CA 94002

or on the Web at

> http://www.thomson.com/duxbury.html.

Chart Wizard for Excel 97

In This Appendix...

- Chart Wizard vocabulary
- Using Chart Wizard
- Resizing and moving charts in a worksheet
- How to rescale a chart axis
- How to retitle a chart
- The Chart Toolbar
- Adding new data to a chart
- Using a chart sheet

The boxplots and histograms studied in Chapter 3 are powerful techniques of visual summary. Each can transform a disorganized chunk of raw data into a picture that reveals the basic character of the numbers: their center, their shape, and their spread. In this chapter we will look at a different set of charting techniques. These work primarily on data that has already been organized into table form, so the purpose of the chart is to reveal relationships among the various table entries, such as relative size or changing magnitide over time. A well constructed chart or graph of this sort can highlight features of the data that might otherwise be difficult to see. They can be easily produced in Excel using a special utility called Chart Wizard. Begin by studying the following table of teen crime data.[1]

---

[1]SOURCE: *Sourcebook of Bureau of Justice Statistics*, U.S. Deptartment of Justice, 1982 and 1991

Teen Arrests for Serious Crime (in thousands)		
	1980	1990
Violent Crime	66.4	68.7
Robbery	34.8	21.5
Burglary	145	150
Larceny	295	303
Auto Theft	41.2	34
Arson	5.2	6

C.1 THE LANGUAGE OF CHARTS

Chart Wizard was used to produce the following column chart from the Excel worksheet displayed beneath it.

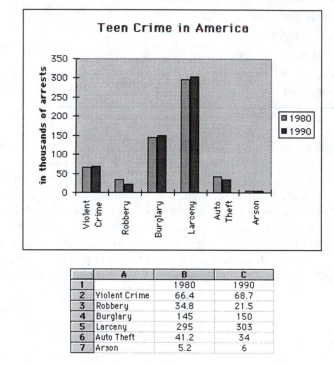

	A	B	C
1		1980	1990
2	Violent Crime	66.4	68.7
3	Robbery	34.8	21.5
4	Burglary	145	150
5	Larceny	295	303
6	Auto Theft	41.2	34
7	Arson	5.2	6

In order to produce such charts confidently, you need to know the language of charts. If you understand what the parts of the chart are called, then you will find it easier to produce the output that you want. If you don't understand this language, you will be at the mercy of Chart Wizard.

The data in the preceding worksheet is organized into three *series*: a list of crimes, data for 1980 and data for 1990. Each data series is in a worksheet column with the *name* of the series, 1980 or 1990, appearing in the top cell. The remaining cells in the data columns, B2:B7 and C2:C7, are called the *values* of the data series. In producing the chart the crime list was used to label the horizontal axis (also called the x-axis). The 1980 data series and the 1990 data series became chart columns, with different-colored bars for each series. (It is important to keep in mind that the phrase "chart columns" refers to the bars of the chart itself and not to the worksheet columns where the original data was typed.) At the right of the chart is a box, showing a color code for each data series. This box is called a *legend*.

Notice that the chart title, "Teen Crime in America," and the vertical or y-axis title, "in thousands of arrests," do not appear in the worksheet. In the section below, you will learn how to attach such titles to a chart. With this vocabulary in mind, let's look at how Chart Wizard works.

C.2 USING CHART WIZARD

1. Open an Excel workbook and copy the teen crime data from **crime stats** in **datach5.xls** into the new workbook.
2. Give the worksheet containing the data the name **Teen Crime**. (Just double-click on the sheet tab and type the name.)
3. Highlight the data—click in cell A1 and drag to cell C7. The result should look something like this:

	A	B	C
1		1980	1990
2	Violent Crime	66.4	68.7
3	Robbery	34.8	21.5
4	Burglary	145	150
5	Larceny	295	303
6	Auto Theft	41.2	34
7	Arson	5.2	6

Make a mental note that the highlighted range is $A$1:$C$7. It will be referred to again.

4. Click on the Chart Wizard button. It is pictured next. Look for it on the Standard Toolbar.

The first step of Chart Wizard will open.

> **Quick Tip**: Use the TAB key to move from line to line in any dialog box. Do not press the ENTER key unless you are ready to go on to the next Chart Wizard step.

Step 1: Chart Type

The Chart Type dialog box is the first of four Chart Wizard steps.

1. Click the tab labeled **Standard Types** to bring forward the set of options shown below.

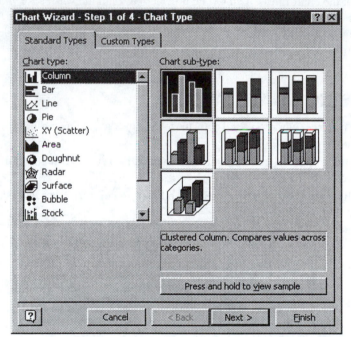

2. Select the chart type you prefer by clicking on its name in the **Chart type** menu. In this case, the column option is already selected. It is the *default* choice, which means that if you make no change, Excel will automatically draw a column chart. Leave the **Chart type** selection set at **Column**.

3. Select the chart sub-type by clicking on its picture in the **Chart sub-type** menu. Here too the default option is just fine, so click the **Next** button.

Step 2: Chart Source Data

The purpose of this dialog box is to make sure that Excel understands where your data is located and how to read it.

1. Click the **Data Range** tab. The dialog box should look like the following illustration.

2. The **Data Range** box contains the location of the data charted, = 'Teen Crime' !$A$1:$C$7. This is the correct range. If it were not, a new range could be entered now before going on. The **Columns** setting for **Series in** is also correct since the crime data for each year was listed in columns B and C of the **Teen Crime** worksheet.

3. Click the **Rows** option to see how this selection changes the chart.

With the Row option checked Excel views the data range as six data series, one for every crime type, with each consisting of an arrest figure for 1980 and an arrest figure for 1990. Thus, the resulting chart becomes two clumps of six columns each.

4. Return **Series in** to **Columns**.
5. Click the Step 2 **Series** tab.

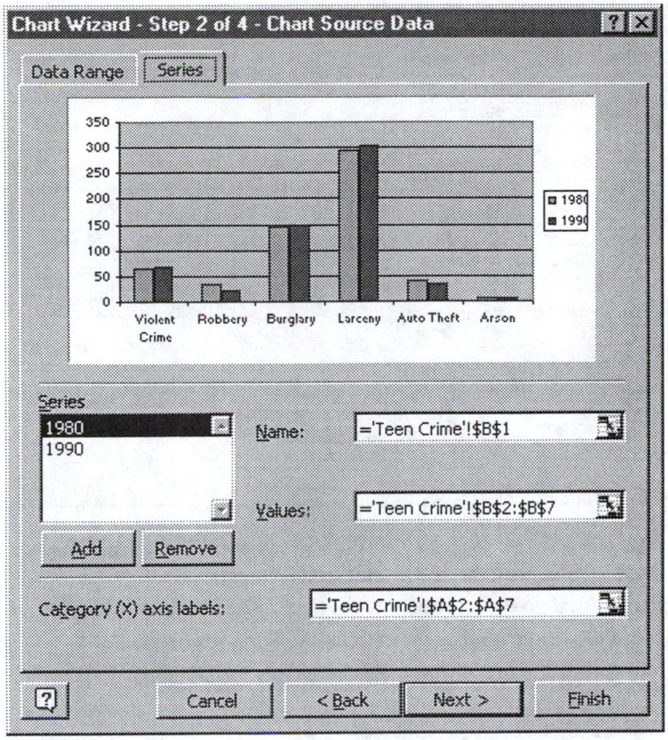

6. In the dialog box above Excel presents its default reading of the columns of the worksheet. Notice that **1980** is selected under the **Series** menu. The **Name** box contains the cell where Excel found the name of this series. The **Values** box contains the range of cells for the 1980 data values. Click on **1990**, and the Name and Values for the 1990 data series will appear in the appropriate boxes. The box labeled **Category(X) axis labels** contains the range where Excel found the labels for the chart's horizontal axis, in this case, the names of the crimes.

7. Since all of the entries in this dialog box are correct, click the Next button.

Step 3: Chart Options

This step consists of a series of dialog boxes which permit adjustments in many chart features.

1. Click on the **Titles** tab and enter the chart titles as shown below.

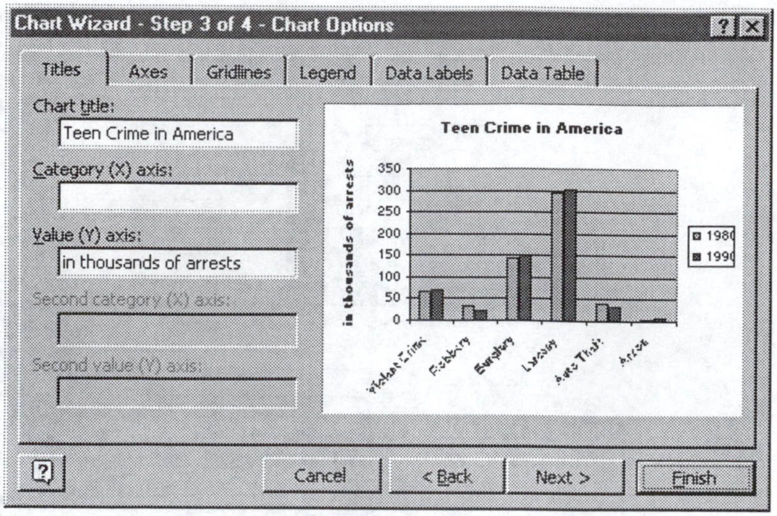

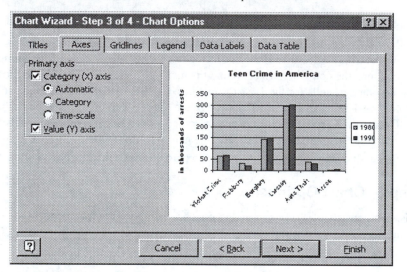

2. Click on the **Axes** tab and the dialog box shown next will come forward. It shows the default settings for the horizonal or *x*-axis and the vertical or *y*-axis of the chart. The options **Category(X) axis** and **Value(Y) axis** are toggles that attach and remove labels from the horizontal and vertical axes of the chart. Click on each to see its effect on the chart, but restore them both to their default positions.

3. Click on the **Gridlines** tab. This dialog box, illustrated next, provides the option of adding or removing gridlines parallel to either axis. Click on the **Value (Y) axis** box labeled **Major gridlines** to remove the gridlines from the chart.

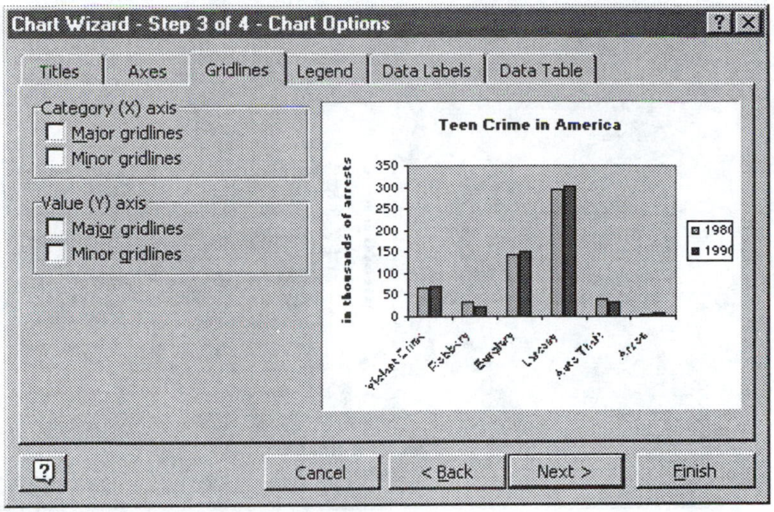

4. Click on the **Legend** tab. Make sure that **Show legend** is checked. Experiment with the various options under **Placement** to see the effect of this option. Click **Right** as shown next before going on.

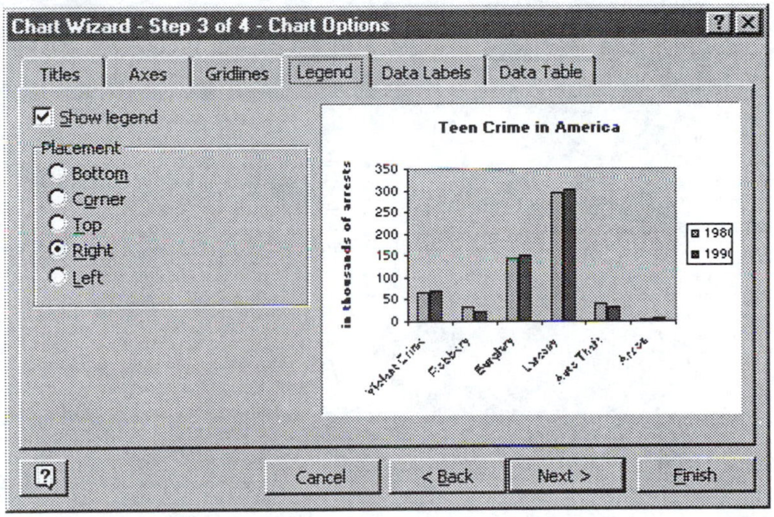

5. Click on the **Data Labels** tab. Experiment with the available options, which place labels on the chart columns, but restore the setting to **None** as shown next.

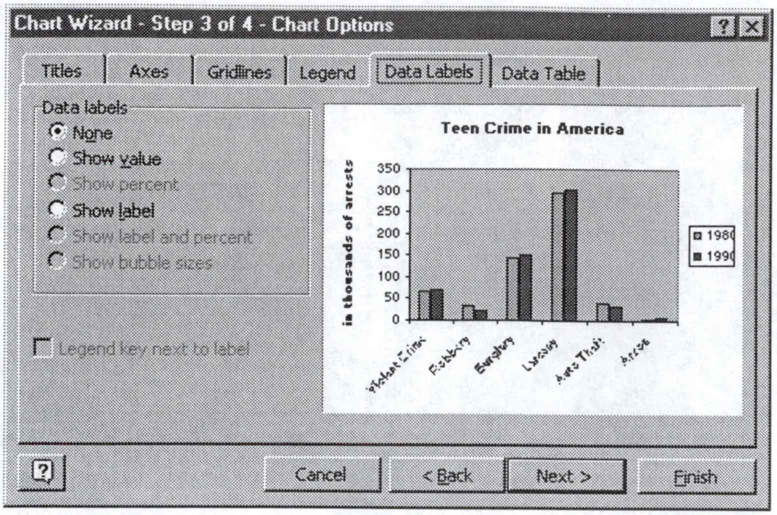

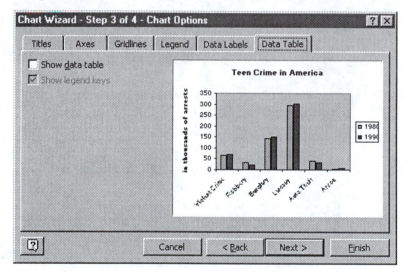

6. Click on the **Data Table** tab. Select **Show data table** just to see what happens. Return to the setting shown next.

7. When all of the chart options are set, click on the Next button.

Step 4: Chart Location

The fourth step of Chart Wizard is pictured below. It offers two options for the final location of the chart. The chart can be placed into a special new sheet called a *chart sheet* or it can be located in one of the exisiting sheets of the workbook. To select the new sheet option click **As new sheet** and type a name for the sheet in the new sheet box. To place the chart in an existing sheet click **As object in** and use the drop-down arrow to select the sheet.

1. Make sure that the Step 4 dialog box looks like the next picture.

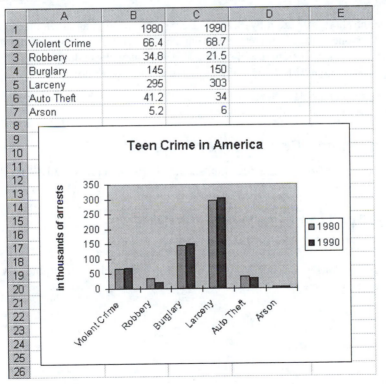

2. Click the Finish button and the chart will be placed in the **Teen Crime** worksheet as shown next.

	A	B	C	D	E
1		1980	1990		
2	Violent Crime	66.4	68.7		
3	Robbery	34.8	21.5		
4	Burglary	145	150		
5	Larceny	295	303		
6	Auto Theft	41.2	34		
7	Arson	5.2	6		

C.3 SOME SIMPLE EDITING TECHNIQUES

You are now ready to put the finishing touches on the chart. You may need to change its size, or, perhaps, you would like to move it to another spot in the worksheet. Maybe you have changed your mind and don't want a chart in the worksheet at all. In this case you need to delete the chart. Each of these actions—resizing, repositioning or deleting—is an editing technique and can be done quite easily once the chart is *activated*. If you are already comfortable with these editing techniques, which were briefly covered in the earlier chapter on boxplots and histograms, skip to Section C.4.

How to Activate a Chart

A chart is *activated* when its border contains eight rectangular handles. Chart Wizard automatically activates any chart as it embeds it in a worksheet. If your chart does not show the handles, click in a corner of the chart and the handles will appear.

How to Change the Size and the Position of a Chart

To resize the chart, activate it by clicking once inside the chart and then place the point of the arrow on one of the eight rectangular handles located on the charts border and click and drag the mouse. The chart will change size and shape depending on the direction of the drag. Experiment with the resizing until you get an idea of how it works.

To reposition a chart place the point of the cursor in the chart border, click and drag the chart to a new position in the worksheet.

How to Delete a Chart from a Worksheet

To remove a chart from a worksheet first activate it and then press the DELETE key.

C.4 HOW TO LIE WITH STATISTICS: MORE EDITING TECHNIQUES

The following chart was drawn by the Wizard from data included in **datach6.xls** in the worksheet titled **crime**.

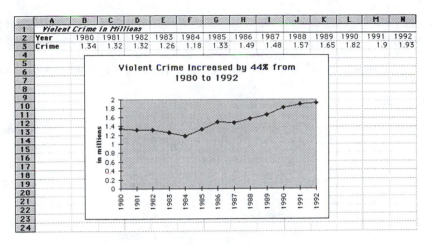

This chart gives a more or less neutral and straightforward presentation of the data. The claim made in the title accurately reflects the facts. Now take a look at the two charts shown below. They were both drawn in Excel from exactly the same crime data. But as you can see by comparing them with the preceding one and with each other, the psychological impacts are quite different.

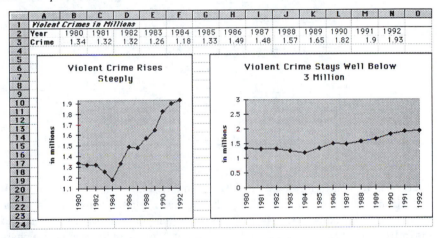

What produces the different effects? The titles, of course, make a difference. We are encouraged in these examples to view the violent crime increases in an extreme fashion, as very large in the left-hand chart and as negligible in the right-hand chart. In addition, the scalings on the vertical axes were adjusted and the charts reshaped to make the pictures seem to support these quite different views of exactly the same data set.

Pictures *can* exaggerate or even lie. In the section below you will learn how to produce charts like the two above. The point of the exercise is this: It is easy to produce charts that misrepresent data, and it is important that you know how this might be done so you can avoid being misled by a bad chart. In addition, there is a deeper problem with all three charts that goes beyond axis scaling or chart titles: none of them takes into account the changes in the population of the United States over the same period. Maybe the increase in crime was simply due to an increase in population. In the exercises (See Chapter 6) you will have the chance to investigate for yourself the effect of this underlying variable.

How to Rescale a Chart Axis

In the chart shown on the first page of the chapter, the vertical axis is calibrated from 0 to 2. Suppose you wish to change this scale to the one shown on the chart titled "Violent Crime Rises Steeply," which runs from 1.1 to 1.93.

1. Double-click on the vertical axis. Rectangles will appear at the top and bottom of the axis and the Format Axis dialog box will appear. (Or, click on the vertical axis once, pull down the Format menu and release on Selected Axis... .)

2. Since you want to scale the selected axis, click on the **Scale** tab.

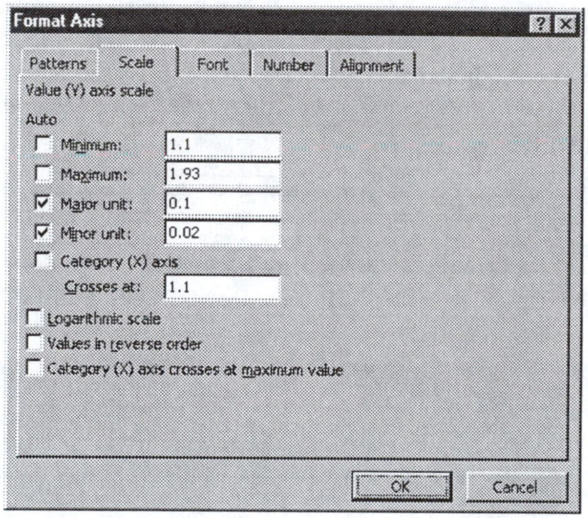

3. Type "1.1" in the **Minimum** box and "1.93" in the **Maximum** box. Leave the **Major** and **Minor units** set as they are. Type the number 1.1 in the **Category (X) axis Crosses** box. Click the OK button. (Remember that you can use the TAB key to move from one line to the next in the dialog box.) You should now have a chart scaled like the "steeply rises" chart shown earlier. Note: you can adjust the length of the horizontal axis, a feature that differs dramatically in the three charts, by simply resizing the chart.

How to Edit a Chart Title

1. Click on the chart's title once or twice slowly (don't double-click) until the insertion point appears in the text of the title. Highlight the title you wish to change by clicking at the beginning of it and dragging to the end.

2. Type in the new title.
3. If you are done, click outside the chart to deactivate it.

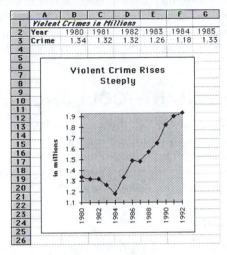

How to Add a Title to a Chart

Notice that in the chart above the horizontal axis is not named. Suppose you decide to add the title "years" to this axis.

1. Activate the chart.
2. Pull down the Chart menu and select Chart Options... .
3. When the Titles dialog box appears type "years" into the **Category (X) axis** box.
4. Click the OK button.
5. Deactivate the chart, and it should now look like the one shown next.

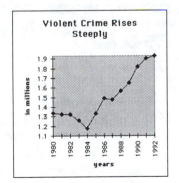

The Chart Menu

The first four options listed on the chart menu, which is visible when a chart is activated, correspond to the four steps of the Chart Wizard. Selecting one of these items will open the dialog box asssociated with the step. Once the box is open, adjustments can be made to the chart just as in Chart Wizard.

C.5 USING THE CHART TOOLBAR

The Chart Toolbar is a handy device that permits you to make certain changes in the appearance of a chart very quickly. This is not an essential feature, since all of the editing techniques on the Chart Toolbar can be done without the toolbar; however, the Excel 97 bar is a great convenience, so you might want to spend a few minutes finding out how it works. If not, go on to Section C.6: Adding Data to a Chart.

Viewing the Toolbar

Pull down the View menu and release on Toolbars... . (If the View menu is grayed out, click anywhere in the worksheet and then pull down the View menu.) When the Toolbox dialog box opens, click on Chart and then on OK. The toolbar shown next will appear on your worksheet.

The Chart Toolbar Buttons

1. On the left is the Chart Objects pull-down menu. It allows you to select an element of an activated chart.

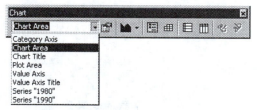

2. The next toolbar item to the right is the Format Selected Object button. It opens the Format dialog box for a previously selected chart element.
3. Click on the drop-down Chart Type button as shown next to display a menu of 18 chart types. This menu is detachable so that it can be dragged to another part of the worksheet.

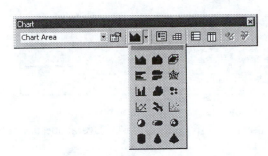

4. Next on the right is a Legend toggle. Click on it to add or delete a legend from an activated chart.

5. Next is the Data Table toggle, which will attach and detach a data table at the base of the chart.

6. The next two buttons can be used to plot the data by row or by column.

7. The last two buttons will angle selected text.

C.6 ADDING DATA TO A CHART

Suppose you come across the violent crime data for 1993 (1,924,190) and want to add this information to the chart.

1. Open the workbook that contains the chart you wish to edit and the worksheet containing the chart data. Type the new data into the worksheet. In the next illustration the data was typed into cells O2 and O3 of the **crime** worksheet.

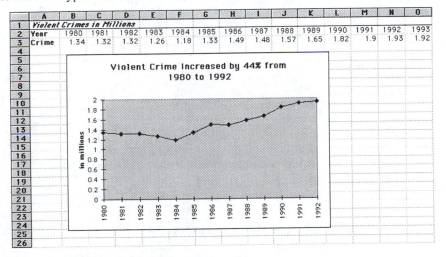

2. Activate the chart by clicking in a corner.

3. Pull down the Chart menu and select Add Data... .

4. When the Add Data dialog box appears, type the new data range into the **Range** box.

Add Data   ⁇ ☒

Select the new data you wish to add to the chart.

Include the cells containing row or column labels
if you want those labels to appear on the chart.

Range: =Crime!$O$2:$O$3

OK

Cancel

---

5. Click the OK button and the Paste Special dialog box appears.

6. The Paste Special dialog box, displayed next, asks several questions about the nature of the data to be added.

Paste Special   ⁇ ☒

Add Cells as
  ○ New Series
  ◉ New Point(s)

Values (Y) in
  ◉ Rows
  ○ Columns

OK

Cancel

☐ Series Names in First Column
☑ Categories (X Labels) in First Row
  ☐ Replace Existing Categories

As the data will create an additional point on the chart, leave the **Add Cells as** option set to **New point(s)**. Similarly, leave **Values (Y) in** set to rows because the population figures, which are plotted on the vertical or y-axis are contained in a row. Make sure that the option **Categories (X Labels) in First Row** is checked.

7. Press the OK button and Excel will adjust the chart to include the new data values.

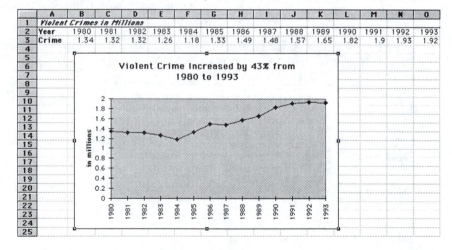

The title should also be changed so that the ending year is 1993 and the increase 43%. $(1.92/1.34 = 1.43)$

Aligning Axis Labels

The labels on a chart's axis are displayed at an angle to the axis. The angle can be changed using the Format Axis dialog box.

1. Double-click on the axis whose labels are to be realigned.
2. When the Format Axis dialog box opens, click on the **Alignment** tab to bring these options forward.
3. The setting shown next, **90 Degrees**, will draw the labels perpendicular to the selected axis.

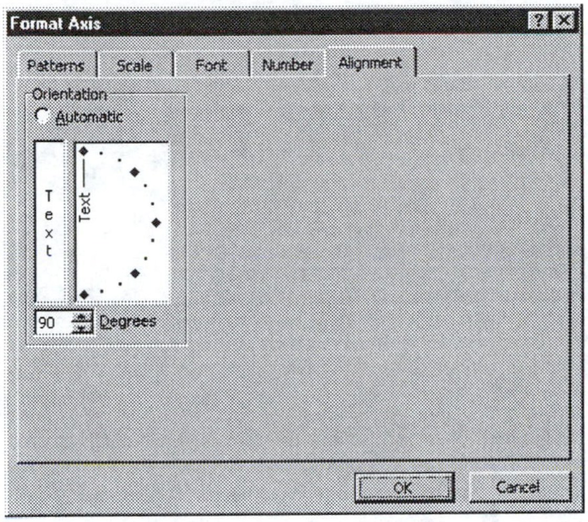

4. The angle of alignment is changed by clicking on the **Degrees** arrows to adjust the numerical value of the angle up or down. The alignment can also be changed by dragging the text dial around to the desired setting.

C.7 CREATING A CHART IN A CHART SHEET

So far all of the charts we have created have been embedded in a worksheet. Sometimes it is more convenient to produce a chart on a separate chart sheet. In this section you will learn how this is done. Open the worksheet **crime and population** from **datach6.xls**.

	A	B	C	D	E	F	G	H	I	J	K	L	M	N	O
1	*U.S. Population and Violent Crime*														
2	Year	1980	1981	1982	1983	1984	1985	1986	1987	1988	1989	1990	1991	1992	1993
3	U.S. Population	225	229	231	234	236	239	241	243	246	248	249	252	255	258
4	Violent Crime	1.34	1.32	1.32	1.26	1.18	1.33	1.49	1.48	1.57	1.65	1.82	1.9	1.93	1.92

Using Chart Wizard

Use Chart Wizard to plot the population data and place the chart in a separate chart sheet.

1. Select the range B2:O3.
2. Launch Chart Wizard.
3. Make choices to produce a scatterplot of the data.
4. In Step 4 select **As new sheet** and type in a name for the chart sheet, say, **pop chart**.
5. Click **Finish**. The chart will be drawn in a sheet as shown next.

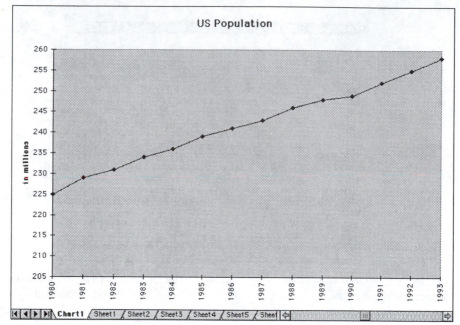

This chart sheet can be renamed or deleted just like a worksheet. You can move from the chart sheet to any other sheet in the workbook by clicking on the appropriate tab.

Copying an Embedded Chart to a Chart Sheet

Sometimes you will want to move a chart that you created and embedded in a worksheet to a separate chart sheet. You might do this if, for example, you wanted to use the chart to make a slide or overhead transparency.

1. Press the F11 key and a chart sheet with a bar chart drawn on it will open. Activate the bar chart by clicking in a corner and press the DELETE key.
2. Click on the tab for the worksheet containing the embedded chart.
3. Activate the embedded chart by clicking in a corner.
4. Pull down the Edit menu and release on Copy.

5. Bring the chart sheet forward by clicking on its tab.
6. Pull down the Edit menu and release on Paste.

EXERCISES

Try the exercises in Chapters 5 and 6. Exercise 5.8 is not appropriately worded for Excel 97 although all of the charts illustrated in it can also be drawn in Excel 97 by making careful choices in Steps 2 and 3 of Chart Wizard.

The Pivot Wizard for Excel 97

D

In This Appendix...

- Using Pivot Wizard to build a contingency table
- Constructing a table of expected values
- =CHITEST

PivotTable Wizard is Excel's table-building utility. It is usually referred to by the shortened title "Pivot Wizard." This appendix contains instructions for drawing a contingency table using Pivot Wizard in Excel 97. Begin by reading Section 15.1 which explains the connection between a contingency table and the worksheet of raw data from which it was derived as well as a few of the technical terms, such as *field* and *category*, that are important in understanding how this Wizard works.

D.1 USING PIVOT WIZARD

Pivot Wizard will be used in this section to construct the contingency table displayed in range A252:D256 from the data in range A3:B249.

	A	B	C	D	E
1	male=1, female=2	Yes = 1, No = 2	right = 1, left = 2		
2	Gender	Smoker?	Handedness		
3	1	2	1		
4	1	2	2		
5	1	2	1		
246	1	2	1		
247	2	1	1		
248	2	2	1		
249	1	2	1		
250					
251					
252	Count of Smoker?	Smoker?			
253	Gender	Yes	No	Grand Total	
254	Male	26	64	90	
255	Female	40	117	157	
256	Grand Total	66	181	247	
257					

Getting Ready

1. Open the workbook **datach15.xls** and click on the tab for the worksheet entitled **gender**. Copy the data to a new workbook and split the screen horizontally so that rows 1 and 249 are visible.

2. Highlight the range of data, A1:C249, as shown next. (Click in cell A3, hold down the SHIFT key, and click in cell C249.) You need not include the headings, rows 1 and 2; Pivot Wizard will automatically include the row 2 labels. (Note: The data on handedness will not be used in this example, but it does not hurt to include it in the highlighted range and doing so will help you to better understand how Pivot Wizard works.)

	A	B	C	D
1	male=1, female=2	Yes = 1, No = 2	right = 1, left = 2	
2	Gender	Smoker?	Handedness	
3	1	2	1	
4	1	2	2	
5	1	2	1	
246	1	2	1	
247	2	1	1	
248	2	2	1	
249	1	2	1	
250				

Step 1 of Pivot Wizard

Pull down the Data menu and select PivotTable Report... . A dialog box containing the first step of Pivot Wizard will open. Make sure that **Microsoft Excel list or database** is selected as it is in the following picture. Click on the Next button.

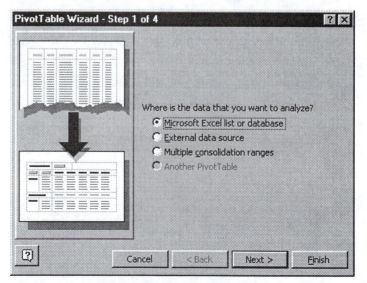

Step 2 of Pivot Wizard

1. Inspect the **Range** box. (It will look like the next illustration.) Notice that Excel has automatically included the row headings in the range. Correct the range entry if necessary.

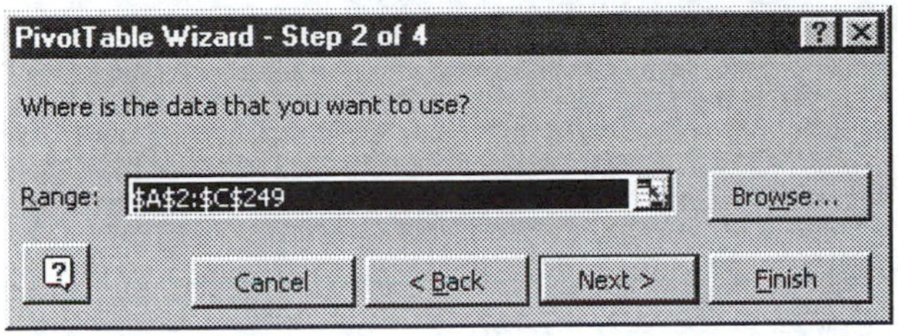

2. Click the Next button.

Step 3 of Pivot Wizard

You have reached the heart of Pivot Wizard. In this step, you actually design the table by telling Excel which fields to use for the rows and columns of the table and by specifying a format for the presentation of the body of the table.

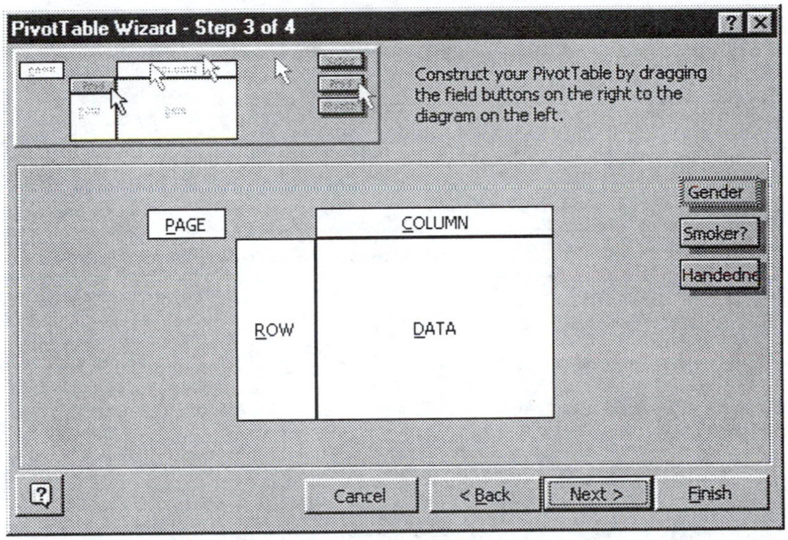

The table you are designing lists the gender categories by rows and the smoker categories by columns. The body of the table consists of *counts* of the number of students that fall into each of the four cells in the body of the table.

1. Click on the **Gender** button and drag it to the word **ROW**. The result should look like the next illustration.

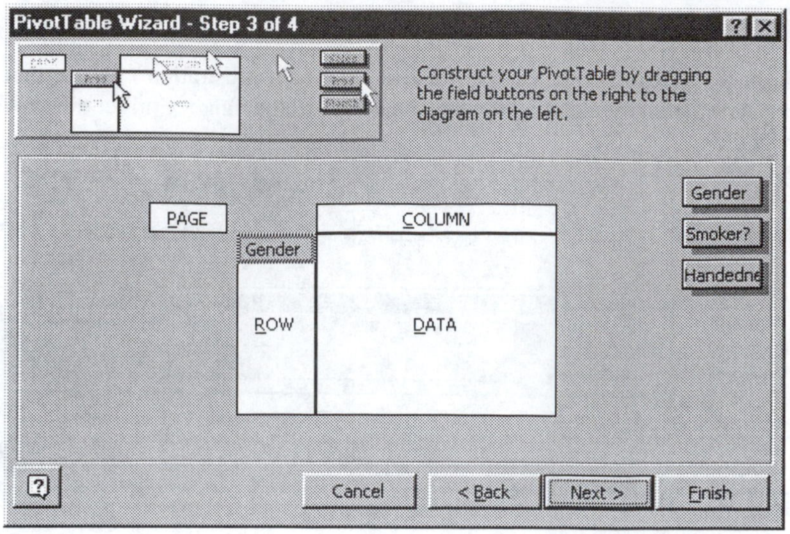

2. Click on the **Smoker?** button and drag it to the word **COLUMN**. The result is shown next.

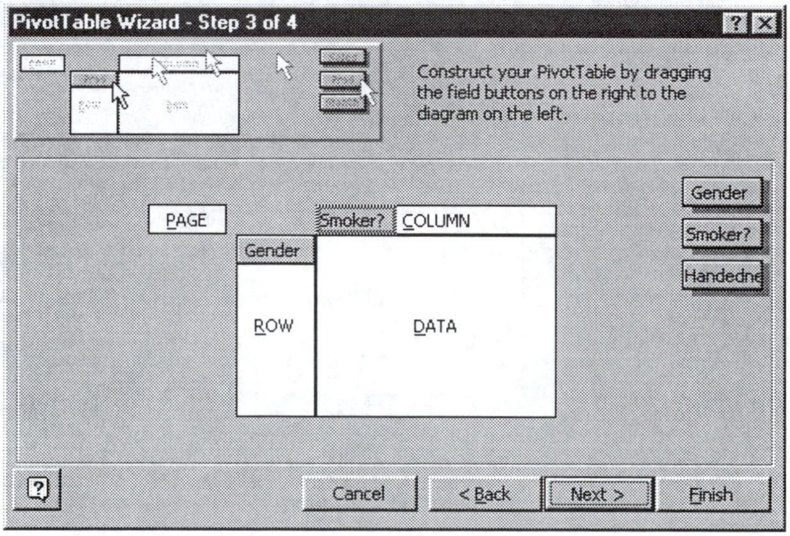

3. Click on either the **Smoker?** button or the **Gender** button and drag it to the word **DATA**. Excel will indicate the default presentation format for the data, **Sum of Smoker?**, as shown next.

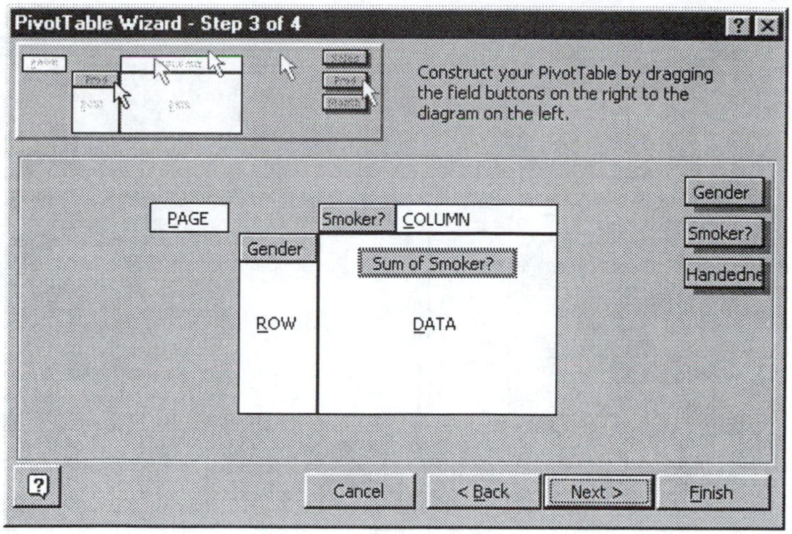

This is not what we want; **Sum** must be changed to **Count**.

4. Double-click on the word **Sum** to open the The PivotTable Field dialog box. When it appears click the word **Count** in the **Summarize by** window as shown next.

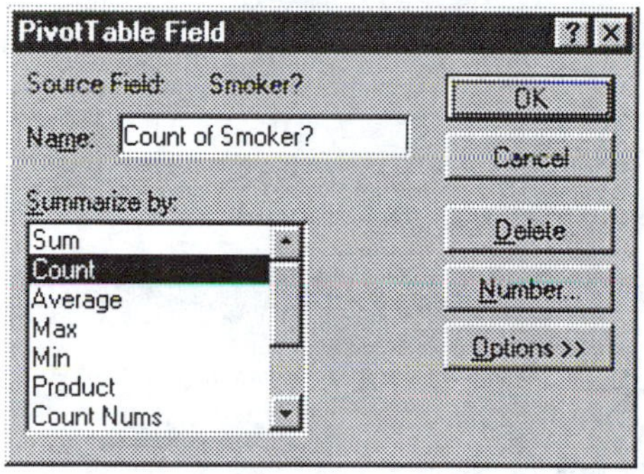

5. Click the OK button. Step 3 of Pivot Wizard will reappear.

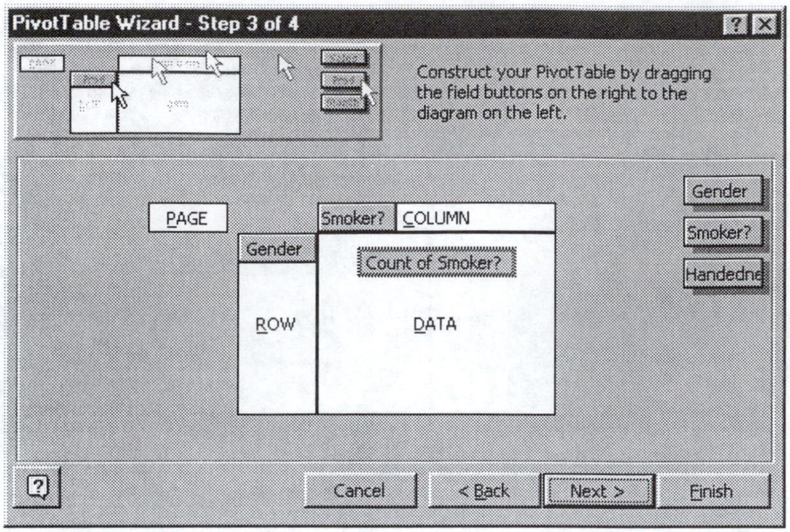

Notice that "Sum" was changed to "Count."

6. Click the Next button.

Step 4 of Pivot Wizard

Indicate the cell in the worksheet where you want the upper left-hand corner of the table located.

1. Select **Existing worksheet**.

2. Type in **A252** as shown next (or click the mouse on cell A252).

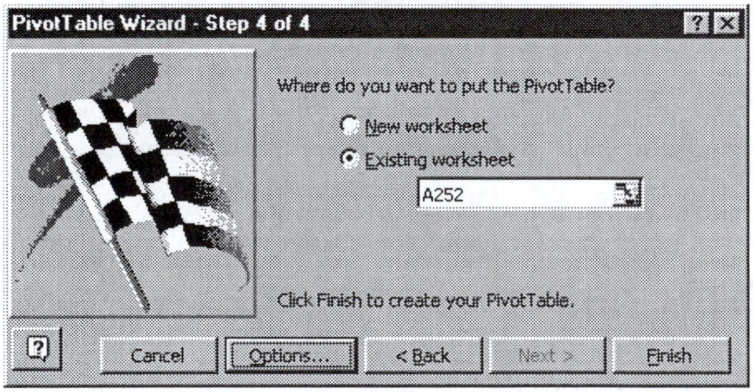

3. Click the Next button and Excel will embed the contingency table in the worksheet. A new toolbar, the Query and Pivot, may also appear. It should be ignored for the time being. If you want it out of your way, just click on its close box.

	A	B	C	D	E
1	male=1, female=2	s = 1, No	t = 1, left = 2		
2	Gender	Smoker?	landedness		
3	1	2	1		
4	1	2	2		
5	1	2	1		
247	2	1	1		
248	2	2	1		
249	1	2	1		
250					
251					
252	Count of Smoker?	Smoker?			
253	Gender	1		2	Grand Total
254	1	26	64	90	
255	2	40	117	157	
256	Grand Total	66	181	247	
257					

D.2 EDITING A TABLE PRODUCED BY PIVOT WIZARD

How to Alter Headings

Some entries in a Pivot Wizard table can be easily changed, but others cannot. Row and column headings are easy to modify. For example, the number 1 under Smoker? can be replaced with the word "Yes" by simply clicking in cell B253 and entering "Yes." The next picture shows the contingency table with renamed headings that make the categories explicit.

	A	B	C	D	E
1	male=1, female=2	Yes = 1, No = 2	right = 1, left = 2		
2	Gender	Smoker?	Handedness		
3	1	2	1		
4	1	2	2		
5	1	2	1		
246	1	2	1		
247	2	1	1		
248	2	2	1		
249	1	2	1		
250					
251					
252	Count of Smoker?	Smoker?			
253	Gender	Yes	No	Grand Total	
254	Male	26	64	90	
255	Female	40	117	157	
256	Grand Total	66	181	247	
257					

Some Changes Cannot Be Made

It is *not* a simple matter to change entries in the body of the table because these counts are linked to the worksheet data entries. Excel will block an attempt to do so with an error message. For example, if you click on the number 26 in cell B254 and then press the

DELETE key, Excel will refuse to erase the value and, instead, return an error message like the one shown next.

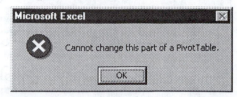

Click the OK button to recover.

An attempt to delete one of the field names, such as "Gender," will produce the same error message.

Formatting Table Entries as Percentages

Some formatting can be altered. For example, the counts given in the body of the table can be changed to percentages of the total number of respondents as follows.

1. Make sure that the PivotTable toolbar, shown next, is visible. If it is not, select it from the View menu under Toolbars... .

2. Click on one cell in the body of the table, such as B254.
3. Click on the **PivotTable Field** button located in the top row of the PivotTable toolbar. The button looks like the next picture.

4. When the PivotTable Field dialog box opens, click on **Options** >>.
5. Click the down arrow for **Show Data As:** and select **% of total**. The dialog box should look something like the next illustration.

6. Click on the OK button, and the category counts will now be shown as percentages of the total number of respondents.

	Count of Smoker?	Smoker?		
252				
253	Gender	Yes	No	Grand Total
254	Male	10.53%	25.91%	36.44%
255	Female	16.19%	47.37%	63.56%
256	Grand Total	26.72%	73.28%	100.00%

For a test of the independence of gender and smoking habits, turn to Section 15.4: An Example of a Chi-Square Test.

$$_nC_r \qquad \binom{n}{r} = \frac{n!}{(n-r)! \cdot r!}$$

Hypergeometric probability Formula

- x non conforming or defective items in sample.
- n number of items in a random sample taken without replacement.
- N ~~lot of~~ number of items in population
- D number of non conforming or defective items in population

HYPGEOMDIST (X, n, D, N)

$$P(x \mid N, D, n) = \frac{\binom{D}{x} \cdot \left(\frac{N-D}{n-x}\right)}{\binom{N}{m}}$$

binomial probability mass function

BINOMDIST (x, n, p) = $\binom{n}{x} p^x (1-p)^{n-x}$ (FALSE)

Binomial Distribution

probability of obtaining x successes in a random sample of n Bernoulli trials where π is the probability of success on a single trial

BINOMDIST (x, n, p) = $\sum_{y=0}^{n} (y, n, p)$ (TRUE) → Cumulative binomial distribution

$$P(x \mid n, \pi) = \binom{n}{x} \cdot \pi^x \cdot (1-\pi)^{n-x}$$

π remains constant during successive trials

BINOMDIST ($x, n, \pi,$ FALSE)

Poisson Distribution

occurrences per unit time or space.

POISSON = $\frac{e^{-\lambda} \lambda^x}{x!}$ (FALSE)

probability of observing x occurrences

CUMPOISSON = $\sum_{k=0}^{p} \frac{e^{-\lambda} \lambda^k}{k!}$ (TRUE)

$$P(x) = \frac{\mu^x e^{-\mu}}{x!}$$

where x = hypothetical number of occurrences in an interval

μ = average number of occurrences in the interval (fraction or int

Table of Excel Functions

Excel function	description	low-tech algorithm	page
=AVERAGE(*data range*)	Computes the average or mean value of a range of data values.	Calculate $$\bar{x} = \frac{1}{n}\sum_{i=1}^{n} x_i$$	42
=CHITEST(*observed contingency table, expected contingency table*)	Computes the p-value for the χ^2 statistic of a pair of $r \times c$ contingency tables, one the table of observed values and one the table of expected values.	Calculate $$\chi^2 = \sum_{i=1}^{rc} \frac{(O_i - E_i)^2}{E_i}$$ and use probability table.	220
=CORREL(*range for x, range for y*)	Calculates the coefficient of correlation of the two data ranges.	$$r = \frac{\sum_{i=1}^{n}(x_i - \bar{x})(y_i - \bar{y})}{\sqrt{\sum_{i=1}^{n}(x_i - \bar{x})^2 \sum_{i=1}^{n}(y_i - \bar{y})^2}}$$	116
=COUNT(*data range*)	Counts the number of entries in the data range.	one, two, three... .	38
=COUNTIF(*range, criteria*)	Counts the number of entries in the range which satisfy the criteria.	one, two, three... .	38
=IF(*condition,display if true,display if false*)	Excel returns *display if true* if the condition is met, otherwise it returns *display if false*.	Sort the data to extract items satisfying the condition.	42
=MATCH(*lookup value,range,0*)	Excel returns the relative position in the range where an exact match to lookup value was found. When 1 or −1 is used as the third argument, a near match is returned when found.	Search the data to find first instance of the look-up value.	187
=MAX(*range*)	Excel returns the maximum value found in the range.	Sort the data until largest value found.	48
=MEDIAN(*range*)	Excel returns the middle value in the range.	Sort the data until middle value is found.	46

Excel function	description	low-tech algorithm	page
=MIN(*range*)	Excel returns the minimum value found in the range.	Sort the data until smallest value is found.	48
=MODE(*range*)	Excel returns the most frequently occurring value found in the range.	Sort the data until the most frequently occurring value is found.	50
=NORMDIST(*x*,*mean*, standard deviation*,0)	Returns the height of the normal curve above *x*.	$\text{height} = \frac{1}{\sqrt{2\pi}\sigma}e^{-(\frac{(x-\mu)^2}{2\sigma^2})}$	139
=NORMDIST(*x*,*mean*, standard deviation*,1)	Returns the area under the normal curve left of *x*.	Calculate $$z = \frac{x - \bar{x}}{s}$$ and use *z*-table.	139
=NORMINV(*percentile*, mean,standard deviation*)	For a normal variable with a given mean and standard deviation, returns the value of the variable having the given percentile.	Use *z*-tables to find *z*-value corresponding to given percentile. Calculate $$x = sz + \bar{x}$$	145
=NORMSDIST(*z*)	Returns the area under the standard normal curve left of *z*.	Use *z*-tables.	149
=NORMSINV(*percentile*)	For a standard normal variable, returns the value of the variable having the given percentile.	Use *z*-tables.	149
=PERCENTILE(*range*,*k*)	Returns the *k*-th percentile value for the data in *range*.	Order the data and assign percentile values.	51
=QUARTILE(*range*, *k*)	Returns the *k*-th quartile value for the data in *range*.	Order the data and assign quartile values.	48
=RAND()	Returns a number selected randomly from the interval, $0 \le x \le 1$.	Use random number table.	128
=RANDBETWEEN(*integer1*,*integer2*)	Returns an integer selected at random from the interval from *integer1* to *integer2*.	Use random number table.	163

Excel function	description	low-tech algorithm	page
=STDEV(*data range*)	Returns the standard deviation of the values in *data range* assuming they are a sample drawn from a population. Use STDEVP for the standard deviation of a population.	$$s = \sqrt{\frac{n \sum_{i=1}^{n} x_i^2 - (\sum_{i=1}^{n} x_i)^2}{n(n-1)}}$$	150
=TDIST(*x,degrees of freedom*,1)	Returns the one-tailed *p*-value associated with *x*.	Calculate $$t = \frac{x - \bar{x}}{s}$$ Use *t*-tables.	206
=TDIST(*x,degrees of freedom*,2)	Returns the two-tailed *p*-value associated with *x*.	Calculate $$t = \frac{x - \bar{x}}{s}$$ Use *t*-tables.	206

Index